LOCK, STOCK, AND BOARDS

THE HARRIS PINE MILLS STORY

To order additional copies of *Lock, Stock, and Boards,*
by Marilyn Nagele Applegate, **call 1-800-765-6955.**

Visit us at **www.reviewandherald.com** for information
on other Review and Herald® products.

LOCK, STOCK, AND BOARDS

THE HARRIS PINE MILLS STORY

MARILYN NAGELE APPLEGATE

REVIEW AND HERALD® PUBLISHING ASSOCIATION
Since 1861 | www.reviewandherald.com

Published by Review and Herald® Publishing Association, Hagerstown, MD 21741-1119

Review and Herald® titles may be purchased in bulk for educational, business, fund-raising, or sales promotional use. For information, e-mail SpecialMarkets@reviewandherald.com.

The Review and Herald® Publishing Association publishes biblically based materials for spiritual, physical, and mental growth and Christian discipleship.

The author assumes full responsibility for the accuracy of all facts and quotations as cited in this book.

Unless otherwise noted, Bible texts are from the King James Version.

This book was
Edited by Penny Estes Wheeler
Copyedited by Vesna Mirkovich
Designed by Emily Ford
Cover art by Daniel Añez
Typeset: 11/13 Minion Pro

PRINTED IN U.S.A.

18 17 16 15 14 5 4 3 2 1

Library of Congress Cataloging-in-Publication Data
Applegate, Marilyn Nagele, 1940-
Lock, stock, and boards : the Harris Pine Mills story / Marilyn Nagele Applegate.
pages cm
ISBN 978-0-8280-2722-9
1. Harris Pine Mills--History. 2. Lumber trade--Oregon--History. 3. Furniture industry and trade--Oregon--History. 4. Harris, Clyde, 1890- I. Title.
HD9759.H36A67 2013
338.7'674009795--dc23
2013015352

ISBN 978-0-8280-2722-9

DEDICATION

Dedicated to all the workers at Harris Pine Mills
who bent into their tasks day after day,
applying skills, genius, and loyalty
with a certainty they were in the Lord's work.

ACKNOWLEDGMENTS

My gratitude goes in many directions in creating this book. First to Jeannette Johnson, acquisitions editor, who encountered me perusing a Review and Herald book table at the 2005 General Conference session; she stacking books, me chatting, whereupon I realized I was talking to a real live editor and made a pitch for this book.

Deep appreciation to the following people:

- Rodney, my husband, who clarified and added to the manuscript and always believed in this project even when, at times, I had given up.
- My dear friend Sandra Ratliff, who endured my first draft all the way to California, she driving, and unable to escape, and miraculously stayed with me during the ensuing years, giving unflagging editing and moral support.
- Margaret De Romanett and daughter Dr. Linda, who handed over family heirloom pictures in perfect trust that I'd do justice to this undertaking in memory of Aunty and Uncle.
- Marcene Garriott, a joy to work with, a member of the Harris clan willing to reveal even the touchy subjects.
- Charlie Fry, Ray Watson, Bessie Freeman McGuire, Lilly Hathaway, Dale Jenner, Gib and Gloria Garver, Billy Tapley, and Donna Bain, who told me stories that turned into chapters.
- Clyde Iverson, who unceasingly added details when I unexpectedly put him on the spot for memories.
- Larry Hersey, who handed over a collection of HPM lore he had rescued from oblivion.
- And of course, bankruptcy trustee John Mitchell, who brought about an ending that satisfied every last creditor and helped save the dignity of the church and the mill.
- Vesna Mirkovich, a copy editor every writer should have.
- Editor Tim Lale, who spoke encouraging words at the very end of this undertaking.
- And to Dwain Esmond, the last great voice of support and correction I heard from RHPA before I finally let go of the project and hit the SEND button.

CONTENTS

PREFACE

Once upon a time a very rich man gave his empire of forests, his sawmill, and his assembly plants to the Seventh-day Adventist Church. It was worth millions of dollars, a fortune indeed in those days. His wife wasn't sure it was a good idea, and his employees thought he had lost his mind.

The story seems like a fairy tale, made more compelling when the news media caught on and came running. *Time* magazine insinuated that $5 million was a lot of tithe for a man to pay to his church, although they had undervalued the gift by much more than half. It asserted that preachers were now running the business and, by inference, Mr. and Mrs. Harris and the Adventist Church were visionaries, not businesspeople. It maintained that despite public demand, the company would not do business on the Sabbath. The stage was set to wait for failure.

This forgotten slice of Adventist history and the risks involved is about individuals who are nobodies in terms of worldly achievement but on page 1 of God's record book. And it is the story of a man who grew an empire with bigger things than money in mind, about how he studied and prayed his way into giving it to his church, and how the church officials took his dream and carried it forward.

It's also about how the church was blessed by this operation.

By the late l940s Harris Pine Mills (HPM) led the world's production of unfinished furniture. The introduction of finished, hardwood, and redwood furniture would come later. But suddenly in 1951 it was an Adventist institution. Clyde and Mary Harris were celebrities in the news and were given rousing acclamation at the 1954 General Conference session, held in San Francisco.

But most of the people who worked at Harris Pine Mills were puzzled. Nothing had changed. Every workday they still streamed through the mill gates minutes before the mournful 7:00 a.m. whistle signaled the start of the day shift, or they arrived at 4:30 p.m. for the night shift. Workers wrestled 16-foot pine and fir into the cutoff saws. They sliced lengths, then sent the slabs down the conveyor belt. I was one of the summer student workers who lined each side of the low conveyer belt, bent on the task of gathering and stacking my assigned size. Summer heat often soared above 100 degrees Fahrenheit. We gulped water to assuage the effect of the voracious sun that beat on the tin roof of the factory. We helped ourselves to the ever-present salt pills at the drinking fountains and endured until the 4:00 p.m. whistle blew.

Winter brought its own unique problem. Then the cold cement floor numbed workers' feet. The tasks at hand didn't feel heroic, and yet inherent in the hard work was the pride the loyal workers took in their individual jobs and skills. Their children gained Christian educations and the 1,000-member church provided a church family. During those years Harris Pine Mills was known for its fair treatment of employees and was on the labor union fair practices list. As branch assembly plants sprouted across the United States, not a few employees were called into management positions.

The church came to Pendleton, Oregon, amid a host of questioning onlookers. Many weren't sure the mill would survive under church leadership. Clyde's and Mary's extended family, some with mill positions, were unimpressed with the Harrises' benevolence. No inheritances for them. Delegations of Seventh-day Adventists heard the news, read the accounts, or listened intently when Clyde and Mary Harris were interviewed at the 1954 General Conference session. Visitors who lined up for mill tours gaped and wondered what could keep men and women tending repetitive machines for years at a time.

I was 11 years old when Clyde Harris made an offer to the church's officials that they couldn't resist. My father was asked to become Clyde's understudy and the manager of Harris Pine Mills. We moved to Pendleton, Oregon, in November 1951. *Life* magazine descended on the Quonset hut gymnasium, interrupting our church service as they assembled their lights and fiddled with f-stops and light meters. Their dangling cigarettes shocked the congregation as they prowled across the rostrum to get the best shots of Clyde Harris, who was scheduled to give the morning prayer.

This story is about good, hardworking people who didn't give a lot of thought to upward mobility. Yet these "salt of the earth" HPM foremen and superintendents were the church's elders, deacons, and Sabbath school teachers. Most of them, keenly aware of their shortage of formal education, had a gracious shyness about them. Their knowledge, wisdom, and innate skills and abilities set them apart from ordinary mortals. They were men of God. Many of them had been enticed to work elsewhere by lucrative offers, but they stayed because they felt a part of this business that did far more than simply saw boards and put furniture together. They worked at a place with a higher calling. Their kids would work there, too, with many financing an education that would place them in the Adventist work.

But that was a long time ago, and many things have happened since then. The business was signed over to the Adventist Church, as *Time* magazine would state, "lock, stock, and boards." It became larger and more prosperous every year. I grew up watching and hearing the stories as they played out. And one sultry August evening I married the sawmill superintendent's son in the new Pendleton Seventh-day Adventist Church on South Hill. And like many, we moved on to further our education and find a place in the gospel work.

What few people know is that Clyde had an unfulfilled dream for his business. His heart was bad. He had no children. But he loved the Lord, and he loved his church. When he was freed from the day-to-day stress of running such a big operation, he traveled the world with Mary. He fished and relaxed at his cabin in the Wallowa Mountains, where he was harassed by game wardens who were certain he was pulling more than his legal limit of fish from the lake and surrounding streams. The Lord blessed his fishing expertise, his health, and his flourishing fortune. Clyde and Mary lived to see their dream of providing means for Adventist education. Harris Pine Mills built assembly plants at many Adventist educational sites around the United States—26 branch plants in all. Clyde loved it. And he never quit thanking God for giving him the dream and bringing it to fruition.

So this is the story. Some of the names have been changed to protect identities or because history did not leave a trail in that regard. But the story is absolutely true. God was in the details of a for-profit business. And just as when God directed Moses to build a sanctuary in the wilderness, craftspeople were needed, and God blessed men and women with special skills. So it was at Harris Pine Mills. Men and women with varied abilities materialized as needs arose. The secular world was nonplussed. How, they marveled, could this business set up and run with a bottom-line objective turn conventional wisdom on its ear time after time? As Clyde told the news media, it was because "it all belongs to the Lord."

The saints in this story are just sinners saved by grace and humbled by the vagaries of life, and that includes the brethren. They all worshipped a God who abounds in a love and kindness that sometimes saints lack. The Harris Pine employees came to regard their jobs as part of the Adventist work of a gospel of grace.

This book isn't so much about Clyde, Mary, Charlie, the brethren, the host of workers, or even the painful bankruptcy saga at the end. Rather, it is about the mysterious ways God works with imperfect people to bless and to save.

MAGGIE'S MIRACLE

CHAPTER 1

It was a gritty, hectic Monday morning, the clang and rumble of Harris Pine Mills in full swing. Just a few months had passed since that fall evening in September 1951 when Clyde and Mary Harris gifted their entire wood industry to the Seventh-day Adventist Church. An August heat wave bore down on the furniture factory's tin roof. By noon it would feel like a furnace inside. Green chain workers, those intrepid men who rhythmically sorted and stacked the water-weighted green lumber fresh from the sawmill, were pouring water over their heads to avoid heat stroke. The black asphalt shimmered around the boxcars where two-man teams loaded trains with finished lumber, braving triple-digit temperatures inside the cars. Your partner shot the boards to you for an hour, then you and he traded places and the outside heat seemed cool by comparison.

In the head office the swamp cooler labored loudly above the squeaky office chairs rolling across the green linoleum, in addition to machines *ka-jinging* and phones ringing. The usual solemnity of typical corporate headquarters did not come to mind when visiting HPM.

Charlie Nagele now sat in Clyde Harris's worn leather office chair. He was still the newcomer on the block since the transition. The General Conference assumed ownership and control of the $15-million operation by December 1, 1951. The knotty-pine corner office was where Clyde Harris had held sway over his empire since moving the Milton Box Company to Pendleton in 1939. It jutted out not 30 feet from the roiling turmoil taking place at the entrance gate. The smudged metal Venetian blinds slanted the eastern sun away in the late morning. But anyone trudging or rolling by at this time of day could glance inside to the boss's desk.

Nagele was on the phone with Dick Denning, the log boss of the woods' operation, and Dick was upset. "The cook quit in a huff this morning," he announced flatly. "No warning at all."

As the transition unfolded, crises were the daily fare. The bosses felt that way, anyway. Taking a for-profit business into a mission-driven church organization called for a delicate balance. Grab the reins but hold them loosely, the brethren advised. It would take time to build employee and business trust.

Now the logging operation was stalled. Loggers were fundamental to the workings of the mill. They had to be housed and fed. Give loggers great food and lots of it and they'd work hard. Pearson Creek logging camp, 40 miles from the main plant at Pendleton, was lined with shacks Clyde had built in a forest of ponderosa pine and Douglas fir. It gave each worker a place to bed down at night on an Army cot with a belly full of hearty chow. The hardworking employees tucked away voluminous amounts of food. Breakfast was the usual—pancakes, hash browns, eggs, steak, sweet rolls, doughnuts, gallons of milk, and coffee, topped off with both number 10 tins of fruit and every fresh fruit in season. That was for starters.

After breakfast the loggers lined up to assemble their sack lunches. Each would slather a loaf of sliced bread with mayonnaise and then heap on the roast beef, tomatoes, cheese, and onions. Maybe they'd put peanut butter and jelly on the two heels and shove it all back in the bread sack, adding a couple of pieces of fried chicken. Then they'd go for all the desserts they could hastily wrap in wax paper, polishing off all the apples and extras they could wedge into the king-size grocery bag. All that would probably hold them until supper.

The food rolling out for these log-bucking employees was spectacular in quantity and quality. At least, that's what was expected. Anything less, and the loggers wouldn't stick around.

"Cook just walked off the job and drove out of here a few minutes ago," Denning added. "We gotta shut down. The workers hafta eat."

Denning had been guardedly skeptical as he watched the church takeover from a distance. It was no secret that as far as he was concerned, Nagele and the church fellows would have to prove themselves. In the tough business of logging, meekness and piety were scoffed at. Did the church have any idea what they were getting into? Loggers didn't suffer fools gladly.

Charlie took a deep breath. In the months since the denomination had assumed control, he had moved to Pendleton from his position as secretary-treasurer of the church's North Pacific Union Conference. He was almost used to tough problems by now, but this news upped the stakes considerably.

"Can you cover the bases for a few weeks until we find a replacement?" he asked as his mind scrambled over the possibilities. "Advertise in the *East Oregonian*, the church periodicals such as the *North Pacific Union Gleaner*, the *Review and Herald*. Get the word out ASAP."

"There's nobody," Denning replied. "His cook's assistants left in a cloud of dust right behind him."

Something was wrong in the woods. For some time rumblings that food and Caterpillar parts were being stolen at the logging camp had drifted to headquarters. The food costs had skyrocketed, so it was obvious that the food inventory was being hauled off for personal use. The walkout that August morning was the result of Denning's confrontation.

"How 'bout scraping by for a while?" Nagele asked.

"Not a chance," Denning replied. "We'll have to shut down until you get a replacement."

"That'll affect the plant and production."

"Yeah, I know."

"Well," Nagele sighed, "it'll probably take some time. I haven't the slightest idea of what to come up with. But I'll call a meeting of the bosses right now to jump-start the process."

The conversation ended. Cooks were hard to come by in the best of circumstances. To find one willing to brave the burly gang out there in the logging camp, well, that was another matter.

It all belongs to the Lord came to Charlie's mind. It was the steadfast declaration Clyde had repeated again and again when he gave the gift of his life's work. It was reminiscent of another statement Clyde used to make, "The seventh day is the Sabbath," which overarched all his decisions regarding the operation's six-day workweek. No arguments could abridge the seventh day. Not the half-dried lumber steadily rolling through the dry kilns, not the powerhouse boiler. Even the loggers knew curfew was sundown Friday night.

Clyde supported Nagele and the Harris Pine Mill team in the transition—from a distance. And in those first months Charlie longed for Clyde's tutelage. The decisions he made were resolute, his energy infectious. Yet the first day that Charlie was in charge, Clyde swept through the plant, Charlie in tow, and, in response to the unspoken but implied desire that he offer continued guidance, said, "You'll never learn how to run this place if I stick around."

Then he disappeared into the Alaskan wilderness. Mary and Clyde went on a well-deserved vacation.

Day after day the mill and team were faced with tough hurdles. Turbulent months pointed to God's intervention. This was no time to get weak in the knees. But a *cook*? This wasn't going to be solved quickly. A shutdown in the middle of logging season foreshadowed a slowing of stockpiling for winter use. Even when the snow was too deep for felling the timber, you had to have logs. Charlie tilted his chair back, looked out the window at the traffic jam at the entrance and the gas pumps, then swiveled as someone knocked on his door.

Jack, the pump attendant, stuck his head in the door. "Mr. Nagele, there's a woman out here who insists she must see you."

"What's she want?"

Charlie held the phone to his ear as he told the switchboard to call an emergency superintendent's meeting.

"She wants a job."

"Tell her to see Bud."

Bud Smith, Clyde's nephew, was the plant boss. Bud had been Clyde's trusted right hand and had grown with the company. He knew that job applicants liked to go around the system with their requests. Bud would take the woman's application and interview her.

"I already told her that," Jack replied. "She says she's not moving until she sees *Elder* Nagele." Jack put the emphasis on "Elder," the title given an ordained minister, a title that did not fit in this industry of logs and commerce. "Hey, we've got a traffic jam out here until we can get her to move."

"Well, tell her that Mr. Nagele says she must move her car."

Five minutes later Jack was back. "Sorry, but she won't take no for an answer. We've tried everything."

Charlie bent the Venetian blinds to see for himself. There was a commotion out there, all right. Clearly the loggers, their rigs backed up at the gate, were exasperated. A weathered Chevy station wagon, riding low on its tires, blocked the gas pumps.

Jack explained to Charlie that he had filled her car with gas and asked her to move her car so the logging trucks could come through for measuring and fuel. But she'd refused. She'd "speak to Elder Nagele first." The ensuing exchange got nowhere. A couple of exasperated loggers pitched in. The old station wagon wasn't moving. Charlie could see that Jack was out of options.

"OK, I'll see what I can do."

The logging camp dilemma was top priority now. But that was the problem with this business. Necessary details and emergencies sometimes collided. How had Clyde managed all those years? *Lord, where do I turn? A cook. We need a cook.* A job-hunting woman clogging the mill entrance was a monkey wrench in the machinery he could do without! But, well, the men had tried.

Nagele pushed open his back door and squinted. Even in the glare, Maggie Sampsel's set jaw and white-knuckled clutch at the steering wheel of her overheated station wagon was obvious. Bumper stickers proclaimed that the driver loved Jesus, thought smoking was bad, and inquired if you were saved. Wild gray hair framed the woman's sweaty, red face. Her generous cotton dress hung loosely on her ample body. Beside her sat a teenage boy, maybe 18. Charlie chuckled. Problems of running the mill aside, this job was colorful. A woman in a calico dress holding loggers and mill men at bay! Whoever she was, she had spirit.

He stuck his head inside the old car and said, "I'm Mr. Nagele. Now, what's all this fuss about?"

Her face lit up. "Oh, Elder Nagele! It's about time these fellers out here know I mean business. I'm Maggie Sampsel, and this here is my boy, Harvey. He's my right-hand man."

"Well, Maggie, what can I do for you?"

Oblivious to the fact this wasn't regular job-hunting decorum, she plunged on. "I read in the *Gleaner* all about Mr. and Mrs. Harris giving Harris Pine Mills to the church. "It's just a wonderful thing," she beamed. "And I need a job. Real bad."

"Mr. Smith does the hiring."

Bud Smith, Clyde's nephew, knew the business inside and out. He'd do the right thing.

"Now, Mrs. Sampsel, if you'll just move your car out of the way, we'll help you make an appointment to see Mr. Smith."

Maggie wiped her sweaty face. "But you don't understand. I need a job *now*."

"I'm sorry, Maggie. "This is summertime. We're not hiring right now. The box factory is full of summer students. But Mr. Smith might have an opening for you this fall."

Maggie brooded. Was she pouting or just absorbing the unwelcome facts?

Charlie took a deep breath. The woman wasn't through.

With a stern look her face leaned in toward his. Voice rising, she scolded, "Elder Nagele, I've got to make you understand. The Lord told me to come here today. So I did, right away, as soon as I could drive up from Portland." She shot him an exasperated look. "You see, Elder," she said, "I'm a cook."

She stared straight ahead in disgust as if the Harris Pine Mill people were a thickheaded bunch.

"A cook?"

Charlie straightened up and stared at the station wagon. It was clear most of her belongings were stuffed inside, including a collection of old aluminum pots that barely cleared the roof. It took him a moment. How often does a cook show up at a wood products operation?

Maggie rattled off her previous work experiences at camp meetings, junior camps, and short-order houses.

"Think you can handle a rowdy logging camp?" Nagele asked, gathering his wits.

"Sure!" Maggie enthused. "Tell me how to get there. I'll handle 'em. Like I said, we came ready to get to work. Everything we need is in the back. Show me where to go."

Charlie reached in his pocket, pulled out an old envelope, and sketched a map to the logging camp on the back. He told her how to get to Pilot Rock, how to recognize the turnoff onto the gravel logging road Clyde Harris had built for his operation, how many miles, and how to recognize the gate to the place. Then he gave her shoulder a pat and handed her his scribbled instructions.

"Maggie, you're an answer to prayer," he said, earnestly.

Maggie glowed, "Isn't God good?" She leaned out the window. "I woke up this morning knowing I had to come to Pendleton ready to go to work. Just like that."

The mill guys at the pump heard the exchange and gaped in wonderment. She shot them her biggest smile and assured Elder Nagele she was up to handling anything. Then, with her weathered arm waving grandly in the air, she barreled out the gate spinning gravel and billowing dust. Maggie had succeeded in getting through to a tough bunch that she wasn't moving until she completed the Lord's instructions to her for the day.

Charlie regarded the nonplussed men around the gated entrance. "Well, I got her to move her car, didn't I?" he chuckled as he sent up a grateful prayer of thanks while striding back to his office. Explanations would come later. As Maggie's car vaporized into the heat, he could only marvel that God works in logging camps, too.

He reached for the phone to call Pearson Creek. "Dick, this is Charlie. I've got you a cook."

The phone went silent.

"Dick, are you there?"

"Ya, I'm here. What did you just say?"

"A cook. I've got you one. She's in her mid-40s, lots of experience. And she's got her teenage boy with her to do the odd jobs."

"But it's only been 30 minutes!"

"Not bad, huh?" Charlie chuckled again at the implausible encounter. "I'll fill you in later. She's on her way and probably will arrive in three hours. Give 'em sandwiches for supper. She'll be ready for breakfast by 6:00 tomorrow morning."

"Charlie," Dick said gratefully, "thanks."

"Don't thank me, Dick. I had help from upstairs."

He put the phone back on its cradle. A need. A prayer. An answer. It was so seamless—as if it didn't happen. But it did. Doubts? Impossible situations? Well, there would be plenty down the road.

Maggie bent into her new job with verve and stayed for years. Loggers could put nothing past her, and once she set her mind she wouldn't budge on principle. All the thievery and chicanery stopped when Maggie blew into camp. The loggers learned to love the blustery woman who wouldn't take no for an answer and refused to allow any shenanigans in her domain. You can be sure she recounted her miraculous tale to the logging crew when she arrived. Harris Pine Mills never hired a better cook or one so faithful.

The rest of the day rolled on without a hitch. The usual commotion reigned outside as trucks ground their gears and air brakes whooshed. At quitting time Charlie sat at his desk reflecting on this day; the Lord had provided a cook. He pictured the persistent, unlikely woman at the pump sitting there in her car and applying for a job. Picking up his keys, he walked out the door, pausing to reflect how he had arrived at this sprawling industry and the day in 1951 when it all began.

THE BRETHREN CLEAR THEIR THROATS

CHAPTER 2

On the evening of September 15, 1951, Clyde and Mary Harris arranged their living room chairs to accommodate a delegation of men. Dark-suited and somber, each clutching a satchel reminiscent of Bugsy Malone's violin case, they climbed the 12 steps to the Harrises' front door. It was unprecedented business that brought these men to Pendleton, the little town in eastern Oregon with the Wild West reputation. The brethren, Seventh-day Adventist leaders, carried their travel-worn briefcases, prepared for an evening of negotiation.

Though it was a comfortable fall evening, the visitors shifted in Mary's French baroque chairs as if they were wearing hair shirts, those garments worn by self-punishing Christians for centuries. The denomination depended on their wise, godly judgment. Yet this evening's vote would throw a wild card into the church's machinery. Not one of these men had a clue how to run a logging operation. Neither did they know anything about running a sawmill or a furniture factory, and the far-reaching distribution lines of the mill. Every year Harris Pine Mills rolled out enough furniture to supply Montgomery Ward, Sears, and a host of other companies across the United States. And lumber? A 1958 General Conference bulletin would report the four-year total of "sufficient dimensional lumber to build 16,500 houses to house a city of approximately 82,500 people."

That night the brethren were staring down the barrelhead of taking on the gift—a for-profit business they knew very little about. Was it a wise move?

Clyde and Mary Harris were prepared to sign over their company to the Adventist Church. Even as the somber group neared the vote to either accept or reject it, Clyde exuded enough confidence for the rest of the assembly. No, he didn't have pixie dust in his eyes. These men were sitting in the home of a realist. This businessman's success was built upon keen judgment and decisive action, while his close walk and friendship with the

Lord had prepared him for this moment. Clyde, with Mary's agreement, was offering up his life's work. And just as Hannah had placed little Samuel's hand in Eli's, Clyde and Mary were ready to watch and pray at a distance.

Clyde had been taking on challenges since he was 8 years old. As the youngest son of a large, barely-scraping-by family, he made his own way selling flowers to wealthy Walla Walla women, as well as fresh trout, for a penny, to anyone with pocket change on the Milton, Oregon, sidewalks. He knew the stakes of this venture better than any of the men who sat in his home that evening. The wood products business didn't mollycoddle. Timber auctions, where a slight nod might buy 30 years' worth of logs, called for an intuitive knowledge of future needs. A wrong call could bring the company to ruin.

It had been a year since Clyde Harris had first placed before church leaders his plan to give the mill to the Seventh-day Adventist denomination. For years he had watched the denomination at work and had no doubts about the church's integrity. Early on in their discussions Clyde and the brethren had agreed that a complete inventory should be taken of Harris Pine Mills. The oversight of such a job fell to the young treasurer of the North Pacific Union. That evening 37-year-old Charles Nagele, my father, also sat in the Harrises' living room. On his lap he balanced a stack of inventory papers that had taken him to Pendleton for most of the summer, for he headed a delegation that had completed the task. Now he was prepared to answer questions and misgivings about the worth and profitability of the business. This meeting would change his career and put his dedication to the Lord's work in the stretch.

By 9:00 that evening the vote was unanimous in favor of accepting the gift. The meeting closed with hearty handshakes all around. A little awed by the magnitude of this unprecedented gift in light of their own modest preachers' salaries, the brethren conveyed admiration and gratitude. Nevertheless, they left grateful, buoyed by Clyde's unswerving confidence and Mary's support of her husband. The president of the General Conference, William H. Branson, and his associates, C. L. Torrey, William B. Ochs, W. Adair, all from the church's headquarters in Washington, D.C., and Elder Cord Scriven, from the union office in Portland, Oregon, paused a moment at the Harrises' front entrance to gaze at the sprawling mill in the valley below. Their business was complete. Harris Pine Mills now belonged to the Seventh-day Adventist Church. A few details still had to be ironed out.

One evening, a week later, our family Packard drifted past the white house above the rock wall on Goodwin Avenue as my dad described the meeting that had taken place inside and the good-hearted couple who lived there. I and my 6-year-old brother gaped at the house that hugged the edge of South Hill. Light flooded around tie-back silk drapes. As I looked up at what seemed so magical to me my 11-year-old imagination soared. Crystal chandeliers glittered through front-room picture windows. I saw brocade settees, gilded mirrors, and exotic lamps while my father, Charles Nagele, described the pastel Persian carpets gracing the gleaming hardwood floors. In the corner where the picture windows met, a luminous ebony grand piano splendidly finished the scene as I considered our scratched-up mahogany spinet.

The room was described by Elsie Dickson in a piece called "Random Reflections," published in the Milton *Eagle*, March 18, 1944:

"The Harris home on Goodwin Drive . . . a beautiful home it is. . . . Painted white and with brick entrance, the front door opens on a small hall, leading to a big living room. Here Mrs. Harris used pastel colors, carrying out those in an Imperial Kernan rug (23' x 13'), with a beautiful mosaic design, intricate and graceful. The fireplace is mirror-backed and accessories include crystal candelabra, with prisms; baroque tables and porcelain figurines and lamps—among them a reproduction of a mid-Victorian, rose-painted lamp with its own tilt-top table. Pictures include an oil landscape by the English artist, Spencer."

My father pointed out that Harris Pine Mills was very big—the world's largest producer of unfinished furniture, in fact. I could see that it took up a sizable portion of the valley below. A tall black smokestack belched gray clouds against the evening sky. Beside it a wigwam-shaped furnace glowed red, shooting sparklers into the air. Sawdust, Dad told us, was being conveyed from the factory and sawmill as the night-shift workers stoked it into the furnace.

Dad referred to the meeting in Mr. and Mrs. Harris's living room when the brethren voted to accept the gift. I pictured gray-haired men with beatific smiles. "Brethren," mainly conference officials, had been my father's bosses. They were dignified men who grasped pulpits and preached vigorously on Sabbath mornings—many of them preaching at a different church each week. Daily they handled tough church problems from behind glossy mahogany desks, that is, when they weren't on the road to someplace around the globe. These men assumed the task of oversight for Harris Pine

Mills situated in this "Howdy, Partner" town, where the yearly rodeo and Wild West pageant stirred up memories of the Old West. Now Pendleton was to become our town, too.

There was a dissonance between the proper dark-suited men from the church's headquarters and the situation they were taking on. They were all thrifty clergy who mowed their lawns in their worn-out black suit pants and frayed white shirts. My childhood centered on church meetings, my father sitting on the platform among the brethren. My mother did all the right things that the wives of church employees were supposed to do while Dad did the big, "important" stuff. He returned from his trips, mowed the lawn, and enjoyed Mother's homemade, hand-washed gluten that he'd named "featherless giblets."

While she ironed and folded another week's worth of white cotton shirts, Dad stuffed sermon notes into his briefcase. My brother sat on his dress shoes, arms around his legs, pleading, "Don't go, Daddy," as Dad tried to steer his way through the goodbyes and out the door again. Just like the families of other church workers across the country, we moved often and played by unspoken rules. We children were expected to behave, not fidget in meetings, and to be polite when patted on the head. And we adjusted.

We didn't announce our presence that night. Introductions would come later. Now 11 years old, I considered myself grown-up and secretly vowed not to embarrass my parents. Moves were a part of being in "the work," and getting acquainted with new people and places didn't always go as planned. Early memories, bolstered by retelling, included me. I was 2, supposedly asleep one Sabbath afternoon in the bedroom of my father's new boss and his wife, Elder and Mrs. H. H. Hicks. My mother, nervous at our first invitation to the Nevada-Utah Conference president's home, peeked around the bedroom door to assure herself that all was well. Instead, I sat in the middle of Mrs. Hicks' pink taffeta bedspread with an entire jar of Pond's cold cream slathered on my face, peering guiltily at her from deep within the goo. Elder and Mrs. Hicks, amused, restrained my mortified parents from punishing me, but they reserved the right to repeat the tale over the years.

We soon learned that the Harrises liked kids and company. Don't knock, and use the side-door entrance, was the rule. Visitors were expected to poke their heads into the kitchen and call out, "Hello." Clyde might be at the table indulging in his favorite supper, hunks of homemade whole-wheat bread doused with cream-rich milk. If so, he'd insist that you join him, and he'd even spoon the cream off the top of the glass bottle and plop it into your bowl. Or he might be sitting in the living room in his favorite chair (white leather) under the floor lamp, the stack of "red books"—his treasured Ellen G. White books—on the carpet beside him, and his Bible ever present. A telephone nestled into the carpet on the other side. His phone etiquette was legendary. With a few clipped *yep*'s he'd acknowledged the call and responded briefly. A distinct click in the befuddled caller's ear, then silence, signaled the call was over. He didn't do small talk or waste time on sign-offs.

Mary's kitchen was her control central. The aroma of freshly baked bread cooling on white towels tempted visitors to go no farther than the kitchen table. She reigned as the best cook in the church. Apple pies, her specialty, were compliments of the abundant crops from their orchard in the Milton valley. Milton Box Company, known for producing an abundance of fruit boxes, had burgeoned through the years.

It was in the late 1940s that Clyde first shared with Mary his deeper vision for their company. She would never forget that day, although she never dreamed how that larger picture of contributing to the Lord's work would come about. One afternoon he strode into the kitchen, his boots leaving a trail of mud. She was used to mud and dirt, but the new idea sent her head reeling. He unrolled a large map of the United States across the Formica table, pinned it down with saltshakers, and began marking X's in different states. "Mary, if we put assembly plants near our academies and colleges, we can provide work for young people for their education. I want to put branch plants here, here, here, and here." From then on it became his preoccupation, his grand design for the company.

But the years added up. Clyde's stamina faltered. Again and again he and Mary mulled the possibilities. He explained that none of their extended family had the vision to carry out his aspirations for their company. When he broached the idea of making a gift of their business to the Adventist Church, she was too shocked to answer. "We'll always have the mill to fall back on" had been her comfort through Clyde's business ventures, the economy dips, and their bank account, which to her, even with a thrift that included darning the runs in her hosiery, was "never enough."

Mary stewed. She knew that her husband's health was tenuous. Dr. John deRomanett, niece Margaret's husband, adamantly suggested that Uncle Clyde slow down. It was true. He stuffed mill details into the vest pocket of his Pendleton shirt, made decisions on the run, and kept too many crucial problems tucked into his astute mind, not on paper. Many times Dr. John would wrap the blood pressure cuff on Uncle's arm, frown in silence as the needle gauge hovered beyond 200, shake his head, and restrict Uncle's activity for several days.

But give the mill away?

Mary hoped it might blow over. Perhaps they could turn the business over to a family member. Then again, she was already painfully aware they were childless. Gradually she adopted his proposal.

The full impact of the decision dawned on her the day the brethren filled her living room for the meeting. Clyde had come home from the mill early to take Mary out for a drive. They cruised South Hill. She loved to gaze at the neat bungalows they had built, their first order of business when they relocated Harris Pine. It relieved the severe housing shortage caused when all the new employees moved their families to Pendleton. Then they drove to the mill, past the bustling furniture factory, the screaming sawmill, and over to the dozen diagonally parked yellow logging trucks, HPM stenciled on the doors.

In her ninetieth year, Aunt Mary, as she was known, recalled the touch of sadness in Clyde's voice as he said, "Today all this is ours. Tomorrow it will no longer belong to us."

"It hit me then," she remembered. "This was a final look at what would soon belong to the church."

Ten years later my father, Charlie, stood before the annual Harris Pine Mills board meeting in Washington, D.C., many attendees who were members of the church's Fall Council. For the record, he told the story of that unprecedented September 1951 meeting in the Harrises' living room. The first 10 years since then were just the beginning of the tremendous blessing the denomination would receive from this gift. Charlie brought his listeners up to speed on the health of the industry, so foreign to a religious organization yet of inestimable value in the form of jobs, donations, and making Seventh-day Adventists' distinctive values known in a secular market.

As he scanned the audience, he saw the men who had been present in the Harrises' living room that pivotal September evening. It was like looking out at proud parents. After all, this was a report on how well the gift had succeeded as they heard the details as told by Charlie, the young man who had assumed the responsibility.

"I arrived in Pendleton on May 1, 1951," he began, "to run an inventory of HPM assets. It was at the request of Brother Harris, who had offered his corporation as a gift to the denomination. The task took three months. Hundreds of pages of detailed piece count of every conceivable item, together with cost or present value extension, were considered. The results were the figures of all the assets of HPM.

"I placed the figures before Brother Harris. He took a casual look at the large volumes of inventory and remarked, 'I thought you were going to do it in a week.'

"Then, he remarked, 'Good! It's bigger than I thought.' Almost as an afterthought, he added, 'I'm leaving for Alaska on Sunday. We'll be gone for a month. You go to Washington and have all the details worked out so when I get back we can do business.' In answer to my entreaties to delay his trip, to help with this important work, he shrugged his shoulders. Grabbing his $25 Stetson, he said, 'You'll never learn how to do things if I hang around all the time and do them for you.'

"So the Harrises went on a long-overdue vacation to Alaska on schedule without further explanation. I went home to Portland for conferences with Elder Scriven [president of the North Pacific Union Conference at that time], and then, of course, the milk run, Portland, Pendleton, and Washington D.C., for endless conferences and discussions with Elders Branson, Torey, Ochs, Scriven, and other officers of the General Conference and the many others who would be concerned.

"Then [there was] the General Conference Committee action to accept the gift and the authorization for officers of the General Conference and the North Pacific Union to hold a meeting at Pendleton soon after the Harrises' return from Alaska to effect that transfer of the company to the General Conference.

"That meeting was held in early September at the Harrises' home in Pendleton. It was an evening meeting with Elders Branson, Torrey, Ochs, Scriven, and Adair, Brother Taft, Brother and Sister Harris, Brother Pflugrad [treasurer of Harris Pine Mills], and I present. Brother Harris and Elder Branson made quick work of the formalities, and the gift was made a reality.

"Attention immediately turned to the matter of a General Conference representative to connect with Harris Pine Mills as an understudy to Brother Harris in running the business. Brother Harris had repeatedly told Elder Branson [the General Conference president] that it must be someone who 'was young and didn't know too much, and was willing to learn. Don't get an old man. He will die when I do,' he had said.

"As the chairman indicated, we were open to suggestions. Then Brother Harris announced he had already made his choice. There followed a hasty meeting in the kitchen by Elder Branson, where I witnessed a one-man nominating committee in action.

"Things happened so fast that I'm not real clear on what took place, but my best recollections recall inferences to the effect that I would be well advised to consider moving to Pendleton if future gainful employment was attractive to me. My mouth opened several times, but nothing came out, and nothing was necessary, as Elder Branson stated every word as an accomplished fact. We returned to the living room. It was then that the real cold fact dawned on me. My friends, Torrey, Scriven, and Ochs, who had been my bosses and fellow workers for years and had never failed to help me in time of need, were really in on the plot all the while."

My father had not known that Mr. Harris would take such a bold stand that September evening in Pendleton when he declared that the gift hinged on his accepting the job. Yet General Conference president Branson was a decisive leader. He recognized the steely resolve of the request in Clyde's eyes and was not deterred. He respected his declaration.

It was during this hastily called kitchen discussion that Elder Branson asked that both my parents be present. My mother, a third-generation Adventist who grew up in a family of eight children on a farm in central California, resisted the job offer proposed to them. From the time she completed her nurses' training at St. Helena Sanitarium and married my father, their lives had been intertwined with "the work." Now, this call seemed to signify leaving denominational work. Our summer vacations were wrapped around the camp meeting preaching circuits. In the Northwest we loved the detours through Yellowstone and the Grand Tetons. Now Mother was being asked to give her approval to my father leaving his position in the Seventh-day Adventist Church to run a for-profit business! This was an uncharted request that had to be seriously considered.

Elder Branson recognized my parents' reservations. Branson had a long record of service—years as missionary to Africa and now the Adventist world leader—and he quickly went to work. Sitting at the same Formica kitchen table where some years before Clyde had unrolled his dream to Mary, Elder Branson leaned earnestly toward my mother and father.

As he spoke, a larger vision of God's work unfolded before them, an unusual denominational call to leave preaching and departmental work, but a unique undertaking that would benefit God's work. Their pay would not change. Later people surmised that the position came with huge monetary benefits, but that was false, and my parents just smiled as the rumors circulated. Except for a downward adjustment in their housing allowance, as Pendleton was considered cheaper than the Portland area, the denominational wages remained the same. Harris Pine Mills was considered part of the Adventist work in every way.

The agreement was reached. My father, would become Clyde's understudy, reporting to the General Conference.

Charlie's account continued.

"One thing can be said in defense of Brother Harris and the brethren, and I say this in as modest a way as possible: their choice fulfilled in every way Brother Harris's wishes for a fellow 'who didn't know too much.'

"I was instructed to report for full-time duty November 1. The intervening days were busy ones, winding up the work in Portland, house hunting, and moving. At 10:00 p.m. on October 31 the last of our furniture was piled into a rented house in Pendleton."

While the larger picture took shape, our family followed the moving van out of the moist climes of Portland. Mount Hood's snowy peak gradually receded in the back window, and we snaked our way beside the Columbia River into the golden, undulating wheat fields of eastern Oregon.

The conference moving van was unloaded, and late that first night we got an introduction to how fast news travels in a small town. A woman wearing a bathrobe, her hair in pin curls, tapped at our back door. (Backyards adjoined in the Sherwood Heights development.) She explained that she had a phone call for an Elder Nagele and assumed that he must be connected with the truck emblazoned with the words "Seventh-day Adventist" that had been blocking driveways all day. Jean and Dale Green lived behind us in the housing tract. The Greens, Shumways, Wysongs, and Applegates became neighbors, the mill and church our common ground. But early on it was clear that Clyde Harris had his own plans as to where we should live.

My father's immersion into the mill operation was like a dunk into the murky log pond. He continued his speech to the assembly.

"At 8:00 a.m. I reported on schedule to Brother Harris at the mill office. With the statement 'Well, you made it,' he reached for his hat, and we made a whirlwind tour of the plant. In 30 minutes I was introduced to 10 or 15 foremen superintendents with the blunt statement 'This is Mr.

Nagele. He is the new manager, and you will take orders from him from now on.'

"I was overwhelmed by the strangeness of everything—from surroundings to people. To remember who was who, and who did what, at the moment seemed impossible. I begged and reasoned to no avail. 'You'll never learn if I hang around and do everything for you,' he said."

My father's former office, the North Pacific Union Conference, was a quiet brick building on Hawthorne Street in Portland, Oregon. Its carpeted halls, hallowed tones, and prim secretaries taking dictation in cherry-paneled offices gave a stark contrast to what he experienced now.

Dad's new office sat squarely in the back corner of the weathered white building at the mill's chain-linked and gravel entrance. The old green linoleum floor announced every heavy-booted visitor striding down the narrow hallway looking for the boss. Peterbilt and Harris Pine calendars adorned the knotty pine walls. Yard bosses milled around with their questions. Outside the squeal of gears, the sound of logging trucks gassing up, and workers crawling over the logs measuring loads and shouting numbers took place 30 feet from the corner office. Levi's-clad men and women—long before jeans were a fashion statement—swung black metal lunch buckets as they trudged through the gate and around the corner of the office. With a glance through the Venetian blinds, anyone could see the activity inside the main office.

My father's report continued. "The rest is history—and to say they have been the toughest 10 years of my life is a masterpiece of understatement, but they also have been the most rewarding. Mr. Harris has been a wonderful teacher. I have learned to love him. My father died many years ago, and I have come to know Mr. Harris even better and closer than I have my own father.

"The Lord has been good to us, and all the gloomy predictions of the many apostles of doom have proven false. Mr. Harris's faith in the denomination's ability to encompass the task and succeed has been justified. I am sure that I express the sentiment of all the brethren when I say that we sincerely hope it has done all he expected it would in the 10-year period.

"A quick look in retrospect shows that sales in March 1951 were $408,195, and in March 1961, $1,078,290. Every year of the 10 has been profitable, with growth consistent and the future security a prime consideration in that growth.

"The brethren of the General Conference and the union conference have been understanding and helpful at every turn. In short, it has been a tough, but most wonderful, 10 years. They can best be summed up by saying that from where I sit, it has been one of those great, once-in-a-lifetime experiences that you wouldn't take a million dollars for—or give a nickel for another one just like it."

Harris Pine Mills would build 26 branch plants at our various educational institutions in the United States. For almost 30 years Charlie sat in Clyde Harris's old chair guiding and watching over Harris Pine's developments with surprise and gratitude at the Lord's audacity in the myriad ways He led the growth and progress. It was clear from the very beginning that God was in the details of Harris Pine Mills. The operation called for large faith and a God who delighted in making this generous gift prosper for so many years.

THE KID ENTREPRENEUR

CHAPTER 3

"Flowers for two bits, ma'am?"

The little kid held up a bunch of gracefully nodding peonies for the woman of the house to inspect. His beguiling smile and ragamuffin clothes competed with the bright blooms carefully wrapped in butcher paper and tied with twine. All along the streets of the antebellum mansions in Walla Walla the women couldn't resist the freckled-faced boy, his eager, bright eyes peering out of tousled sandy hair. From his first endeavors, 8-year-old Clyde Heath Harris was a smashing business success. What woman wouldn't want to display the gracefully nodding blossoms on the hand-carved oak table her grandmother had carried on a wagon headed west?

His mother waited in the buggy and steadied the horses in the sun-dappled shade of the poplar trees, watching her son develop his salesmanship. It was worth the 10-mile buggy ride into Walla Walla. She encouraged his interest in growing things and helped him devise a greenhouse behind their house. When the flowers bloomed, she guided him as he picked and trimmed the blooms to sell. Now as he dashed to the buggy for more flowers, he was emboldened to ask his mother, "Do ya think I could double my price?" And his pockets sagged with more quarters. How Bertha loved her eighth child!

Clyde Heath Harris was born in Milton, Oregon, on March 9, 1890, four years after the Harris family journeyed west. His earliest memories were a sun-baked, treeless yard behind the tiny house his family rented. These early impressions of summer's oppressive heat and no shade except the side of the outhouse filled him with a yearning for rivers and lakes and land with trees.

His mother was thin and somewhat frail for having a brood of nine children. But looks were deceiving. Though she always struggled to keep her family fed, she had the heart of a spirited pioneer woman. Watching her youngest son as she rested under the shade trees of the oldest town in the state of Washington, Bertha took pride in this son who was so earnest and completely engrossed in his business. These were her few singular moments of pride as she labored to meet her family's needs.

James, Bertha, and their six children had rolled west from Waupun, Wisconsin, in February 1886 on the Union Pacific Railroad. Within five years Bertha would give birth to three more children, Claude, Clyde, and finally Hazel. As families go, there was more to the story than a family's urge to come west and a mother's fond desire to watch her son succeed.

At various times leading up to the Harrises' journey west, James held jobs as a jockey, carpenter, and merchandise broker, but gambling was his first love. Sometimes Bertha felt it supplanted his family. More dreamer than practically minded, he envisioned running his own business.

So, back in 1886, he decided to build and operate a saloon, by all accounts against Bertha's wishes. It surely wasn't with a nest egg from hard work and thrift. James was gambling that he'd do a land-office business and quickly pay back the money he would borrow in order to build. The saloon interior would be that of a Victorian dance hall with gilded mirrors, heavy velvet curtains, and many gaming tables. Hard liquor would flow freely.

Finally the day came when the saloon was finished and James stood back and smiled. All he had to do was throw wood shavings across the wooden plank flooring and fling open the double doors for business the next day. Toe-tapping music, drinks, and gambling would lure in customers and pay off the big debt. He was about to prove to Bertha that her opinion of his business acumen was dead wrong. Very soon he'd be flush with money and out from under his old gambling debts. He'd pay back the lenders of the new business and show his wife, that at long last he was dependable. Bertha was about to heave a sigh of relief. Finally the family would have a steady income in spite of her misgivings about the new saloon.

That night a sinister flame suddenly flickered at the rear of the saloon. A shadowy figure sprinted undetected into the blackness and obscurity. The flame wrapped around the building, then shot skyward into the frigid winter air. Instantly, it seemed, the whole building was ablaze. Someone sounded an alarm, and throughout the town of Waupun, volunteer firefighters sprang from their beds into stiffened, frozen pants and raced toward the blaze, only to find water pipes frozen and no water.

The blaze was immense. The newly formed fire department of 40 volunteers could do little with the frozen hoses. Sirens and the eerie glow roused sleeping town residents. Many rushed to the scene, but they were helpless as the new saloon went up in smoke.

James despaired as his investment blazed away. Fire insurance didn't exist in 1886. In the gray light of dawn the cold reality was clear. His dream lay smoldering in the stinking ash. His promises to pay the workers and merchants with his handsome profits as soon as the establishment opened its gilded doors were buried in the piles of refuse.

It takes one to know one, the old saw goes. Most likely he had sweet-talked some of his gambling buddies into putting up cash on what he insisted was a sure thing. But the pile of smoldering gaming tables, the rafters now like so many charred pickup sticks poking through the rubble, fueled resentment toward the poor man with the big ideas. His dream had become a nightmare. His bright plans, his expectations, had vaporized into smoke.

James and Bertha reeled at their financial extremity. If loss of the saloon was not tragedy enough, a week later an unquenchable fire burned their house to the ground. The family escaped, but now they counted their fortune in lives saved. Nothing else remained.

For this reason James, Bertha, and their six children, George, Myrtle, Mattie, Clarence Burdette, Jim, and Darrell left Waupun almost immediately on a westbound train. Distance was the only thing James could put between his family and his unpaid debts and revenge-bent creditors.

Six days later, in February 1886, the Harris family climbed off the stifling train onto the platform at Walla Walla, Washington. Father James came first, then Bertha, holding 2-year-old Darrell, then 16-year-old George, with the rest of the children trailing behind. Uncle Higbee, James's affable, eccentric younger brother, met them smiling broadly. He led them to his carriage and on to his small house on the north side of Milton, Oregon, about eight miles from the train depot.

The historical record states that 20 years earlier "Higbee Harris herded sheep for a living as he traveled west in 1866. Harris was only 12 years at the time, having been born in Waupum, Wisconsin, March 26, 1854. He arrived here with only a dollar in his pocket," *The Valley Herald* would note. It went on to state that the young shepherd "met Joe West in the mountains near LaGrande in 1866. The two men, then both bachelors, went into a partnership to homestead land about three miles up the Walla Walla River, and later bought adjoining lands."

Higbee became wealthy with his aggressive land deals to pioneers streaming in with large families. A risk taker of a different kind, he probably staked his brother's family on their trip west.

James and Bertha remained with Higbee for the rest of the winter. But while Bertha appreciated her brother-in-law's hospitality, she was pregnant again and wanted her own place. An added fact could have been that Higbee's housekeeper and cook cared nothing at all about cleanliness, and her cooking so repelled the farmhands that they chose to cook for themselves over an open fire. Higbee would one day marry his housekeeper, and in keeping with his offbeat reputation, one of their daughters took to patrolling her father's property along the Walla Walla River with a shotgun aimed at cousins who fished too close.

The family rented a four-room house on the northern outskirts of Milton, in what the community called "Poorville" or "No Man's Land." Clyde was 5 years old when a check arrived addressed to his father. He would always remember the exact amount, $457—his father's Civil War pension.

"We'll buy a home," Bertha firmly declared. James didn't argue.

So nine years after moving to the valley, they paid $350 for a house with an acre of land flush with apple, prune, and cherry trees. Best of all, the property bordered the Walla Walla River. The contrast between the barren rented house on a dirt lot and the thrill of a place with trees, garden, grass, and a river never left Clyde. It shaped his decisions and taught him careful planning to keep his eyes on what was of utmost importance to him—land, trees, and abundant produce.

Selling flowers wasn't young Clyde's only business venture. By the time he was 8 years old, he was buying his own clothes and other necessities. Fishing, his great sport, became another of his moneymaking businesses. He mastered the nuances of bait, hook, and the careful teasing of his pole. "Clyde can pull fish out of a bucket," his family would say. The Harris family ate a lot of trout. On the town boardwalks he sold them for a penny apiece, already cleaned, by customer demand.

Bertha took in washing from the Miller family. Aaron Miller was already becoming the foremost nursery stock producer in the Northwest. He grafted and sold young fruit trees to farmers and orchardists across the country. The Miller family became Seventh-day Adventists while attending an evangelistic series held by Elders Loughborough and Bordeau in Windsor, California, where they lived at the time. They were baptized along with James and Caroline Woods, who had at one time been Milton pioneers.

It was an interesting turn of events that prompted the family to move back to Milton in 1871. James Woods, Caroline, and children had moved from Milton to Windsor, California, to escape his fanatical Adventist in-laws. James reasoned there were too many Adventists sprouting up in the Walla Walla Valley. He wanted to be far away from such strange theology.

Who knew that one of the 50 Adventists in the whole state of California would be James Woods's neighbor in Windsor, California? William Nichols was not your average neighbor nor your average Adventist. He was stubborn, hardheaded, and outspoken, as Adventist Church leaders in the Northwest would later complain. But he was persuasive. He convinced James and Caroline to attend evangelistic meetings with the explanation that maybe they had missed something about Adventism.

James and Caroline were stirred by the messages, and James, in an all-is-forgiven attitude, moved his family back to Walla Walla. He turned tables on his neighbor in a persuasive sales pitch for the Walla Walla Valley. He convinced not only the Nichols family but also Aaron Miller's family to move north. Both men became very wealthy and were pivotal in Milton church growth and providing a thriving church school for the children in the valley.

In 1880 the Nichols farm hosted a camp meeting that was attended by Ellen White, who stayed with James and Caroline Woods. She delivered words pointed to the Adventist landowners who were becoming wealthy in this pioneer territory. The getting of wealth had turned their heads, she said, and failed altogether to soften their hearts and their independent pioneer ways. She reminded them God was responsible for their successes. Naming names, she pointedly urged the farmers to make good on the pledges they had made before God in inspired moments and had conveniently reneged. She pointed out that some of their crop failures added up to their pledged amounts. She told the well-heeled men that she would personally pay off their pledges if they were still unwilling, because "God wants a cheerful return of tithes and offerings." The chastened men gladly paid up.[1]

Mrs. White's counsel sank deep into the hearts of the Miller family. When Bertha Harris was hired to do the Aaron Miller family's wash after

[1] Doug R. Johnson, *Adventism on the Northwestern Frontier* (Berrien Springs, Mich.: Oronoko Books, 1996), pp. 10, 11.

arriving in 1886, she admired the Millers and their reputation in the community. She listened closely to why they were exemplary, and became a believer. She was baptized, and although she rarely went to church, because of ill health and lack of proper clothing, she often quoted Scripture to her children. "The seventh day is the Sabbath. Remember the Sabbath day," she would say with a smile as she sank into her rocking chair when the Sabbath began.

Bertha taught her son Clyde about paying tithe. "Ten percent of your earnings belong to God," she admonished him. He never forgot. " 'Bring ye all the tithes into the storehouse,' " Bertha would quote to him. " 'I will pour you out a blessing, that there shall not be room enough to receive it.' " Clyde absorbed the instruction and determined to do right by God.

Young Clyde knew little else about his mother's religion, but he always staunchly stood on those two biblical principles. His motto, "You do right by God and He will do right by you," served as a cautionary tale against the selfish grabbing of wealth and failing to look on one's abundance as God's blessings. We might ponder his stated philosophy and argue his was a bargain with his Lord, but Clyde's faith went deeper. While his outward expressions were a means of simplifying what he believed, he was acquainted with a loving God who loves unconditionally.

Clyde was forever "little brother" to his doting siblings and parents, who rarely could withstand his persuasive pleas to aid his schemes. Every day the OR&N train pulled in to the Milton substation. During fruit season he'd be there enticing travelers with his "Cherries for a nickel! Apricots, peaches!" And of course when fall came, he had beautiful Delicious apples. Who could resist? Surely not the travel-weary people longing for refreshment.

"Poorville," the town that sprang up beyond Milton, was the result of disgruntled citizens yearning for the saloons and gambling that was not allowed in Milton, which was known as a temperance town. A case was made for establishing a new town. Then the question arose, What shall we name it? For a time they settled on the name Wallette. But the Milton *Eagle*, the area's newspaper, hooted, "Wallette? They'll get there yet!"

The council compromised. It would be New Walla Walla. But the Walla Walla *Union Bulletin* had a field day with that name. The council debated again. They were advertising free water as an incentive to entice people. Then someone shouted, "That's it! Freewater."

The name stuck, in spite of the fact that scornful Milton residents referred to Freewater as Jerkwater. By the time Clyde was 12 years old, Freewater had six saloons and more springing up. Among the colorful names were The Boozerino and The Palace, which advertised roulette, plus the first drive-through. You could ride your horse into the saloon and pick up your drinks at the counter!

Of course, the Harris boys learned how to gamble. Living close to Freewater, they identified with the intrepid pioneers who stuck out their chins and held their heads high in spite of the derision of the surrounding area. And one day the disharmony died. But it wasn't until the 1950s that the two towns merged with a hyphen in their name, Milton-Freewater.

"I grew up on the wrong side of the tracks," Clyde would remember. "My father was a tinhorn gambler."

In the Western parlance of the smoky back rooms where the drinks were served, the men gathered around tables to win back what they had lost the night before. The tinhorn was the guy with the fancy dress, smooth talk, and loud stories about past big winnings. Yet when he sidled up to the cheapest table and used a tin cup to contain and shake his dice instead of the snazzier leather horn/pouch, gamblers dubbed him a tinhorn.

Although Clyde was born into a gambler's family, he early set out to achieve higher goals than winning at poker. He may have been a gambler at taking risk, but he turned his keen mind toward improving himself, planting and reaping and investing his profits. He was small for his age and a target for the school bullies. But not for long. His quick mind and agile body were more than a match for young tyrants. Every altercation taught him something about life. If he could outwit his opponent, he stood on a vantage point. He learned wily rules of fist fighting, using brains and quick thinking instead of blundering into situations with no plan of action. On the rare occasion that he came out the loser in an altercation, he learned to bide his time. "Just wait. Someday I'll be standing in his path." He would teach his opponents lessons they didn't soon forget.

His mother's knuckles were red and chapped from hard work. Bertha had hoped the move would change James. But even as he took odd jobs and did a little carpentry, his gambling habit continued. It was well known in the valley that the Harris family didn't have enough to eat. Once some good-hearted women took it upon themselves to provide a basket of food and clothes for the needy family. They hadn't taken into consideration Bertha's stouthearted independence. "Thank you," she told the well-meaning church women as she stood blocking the entrance of the overflowing basket they held. "We will not take charity." Bertha, pious and proud, did not rely on handouts. She wouldn't be a project for the town's pity.

Older brother Dette had a natural ability for music. Clyde copied him, and together they traded off playing for pay at saloons and dance halls. What the Harris kids lacked in money they made up for in talents and spunk. In 1939 a Milton *Eagle* newspaper reporter reminisced about growing up with the Harris boys:

"These two boys were Dette and Clyde Harris. No one could ever guess what they were going to do next. Dette, by chance, came into our all-year home, flopped down on the piano stool, by the family heirloom [the piano], swung around, ran his fingers over the keys, and the room was filled with melodious music. Though Dette never took a music lesson, he could play any piece he had ever heard as well as the composer. Though not a piano tuner, he tuned the old Hallet Davis for us, the first for some years, the only piano in Walla Walla."[2]

The old Hallet Davis had once belonged to Caroline Woods, the wife of James Woods, who had moved to Windsor, California, to escape the Adventists. Soon after coming across the continent with her parents, Caroline's father, Stephen Maxson, ordered the instrument for his musical daughter. The rosewood piano was shipped from Boston, went around the southern tip of South America, up the Pacific Ocean to the Columbia River, and by barge to the Wallula landing for the bumpy 40-mile wagon ride to Milton.[3] In 1880 the talented Caroline sang for U.S. president Rutherford B. Hayes on his visit to Walla Walla. The heirloom instrument would provide Dette and Clyde the opportunity to learn ragtime dance tunes.

The gold town of Shaniko, now a ghost town on the high desert of central Oregon, was once a clearing house for bands of sheep and a railroad line from which to ship. It was also a rowdy way station for thirsty and

---

[2] Lee W. Roberts, "Logging of Blalock," Milton *Eagle*, Apr. 20, 1939.

[3] Johnson, p. 9.

entertainment-starved high-desert sheepherders who came to town to gamble and dance. Dette and Clyde traveled down the Columbia River turning south to the high deserts to play at the Shaniko Saloon. But as he approached adulthood, Clyde grew tired of the superficial life of part-time muscian, occasional gambler, and small land dealer. He began a restless search for a new direction.

BOXES GALORE

CHAPTER 4

A wrong address, an offer to make a little summer money, and an interrupted fishing trip added up to serendipity—or providence—for 22-year-old Clyde.

The summer of 1912 Clyde and a fishing buddy struck out on a trip to the streams in the Wallowa Mountains. Clyde had been in a quandary as to what his future held, and the reflective power of fishing always helped clear tumultuous thoughts in his head. The two young men planned an overnight stop in Cove, a tiny fruit-bearing hamlet near the Wallowa Mountains at the edge of the Grande Ronde Valley. Its artesian hot springs served to warm shivering souls in the harsh winters or as a soothing respite to soak a weary body after a hard day of hiking and fishing.

Clyde had made a little money, first buying a half acre of land, planting prune trees and selling his harvest, then buying an acre of land, until this time he owned five acres of prunes. He had learned the art of planting, tending, picking, and marketing his produce.

Arriving in late afternoon, anxious to get a room and some food in their stomachs, Clyde and his friend trudged down to the Cove Hotel, in the center of the tiny town. The place was dark. They squinted into the windows in the early dusk and noticed the sign dangling at the front door: "HOTEL NO LONGER IN OPERATION."

Deflated and hungry, they wondered, *Now what?* "Well, let's ask someone if there's another place in Cove to stay," Clyde suggested. Down the street they noticed a man trimming his bushes and asked where they might find a room for the night. "Mrs. Long's Boardinghouse, the white house up the hill" was the reply. Gazing at their dusty shoes, they knocked at the first white house they came to. A woman answered the door with a warm smile. "Good evening, gentlemen."

"Is this Mrs. Long's Boardinghouse?"

"No, this is the Reed residence. But if you need a place to stay for the night, my husband and I would be glad to put you up. I would guess you're hungry too," she added.

Sometimes it's the unexpected events or ordinary things that add up to the pivotal sparks that set a course for life. The providence of the wrong address would set Clyde on a path to his future.

Mr. Reed happened to be foreman of a wholesale fruit-distributing business. Clyde talked far into the night with his new friend, learning about growing fruit and the distribution business. Reed liked this clear-eyed, bright young man who was quick to learn. By the end of the evening Clyde had agreed to work for Reed, filling in for a sick employee. The fishing trip was put on hold. Clyde would be staying awhile.

Reed took ill some weeks later, and Clyde acted as foreman while Mr. Reed convalesced. During this time Clyde noticed that next door, about 75 yards from the distributing plant, was a small boxmaking business. A man named Al stayed busy hammering box shook (lumber cut to size for box manufacture) together to sell to the fruit-shipping operation that Clyde now worked for. Clyde made friends with the amiable boxmaker, plying him with questions about the boxmaking business. He also convinced the man to give Reed and him a better price for the boxes he sold to Reed's distribution company.

Late in the season as they were hanging out after work shooting the breeze, Al said, "I've made $4,000 this summer."

Clyde was suddenly all ears. With a low whistle he replied, "Compared with this foreman's job, that's a lot of money for a summer's work." His mind began to mull a new idea. Could he, should he, consider the box-making business too? How could he cut costs and produce boxes for less? His mind roiled with questions.

"How much would I need to start my own boxmaking business over in Milton?" he asked.

"Oh," Al replied, rubbing his chin thoughtfully, "probably around a thousand dollars."

"How soon could I realize a profit from that kind of business?"

"Well, the farmers around here brought their business to me right off, so I didn't have a loss from the first," Al told him.

Clyde would remember that conversation when he plunged into box making in his region. Fact was, at first the Walla Walla Valley farmers didn't line up to buy his boxes, and he would learn soon enough that Al had grossly underestimated the start-up costs of such a business. But he felt he was on to something. And as the acting foreman of the Cove Fruit Packing and Shipping operation, he had the authority to offer jobs to his brothers. He quickly penned a letter to them saying, "Come over here and work for me. The pay is good, and I have an idea for a business, but I'll need your help."

All summer the loads of berries, cherries, peaches, and pears rolled in, and in early fall it was the beautiful Red and Yellow Delicious apples. Clyde never regretted canceling his fishing trip. Boxing and shipping produce with Al as his coach, he surged with new energy as he envisioned the future.

Milton Box Company was the logical name for their new business. But first they needed money and knowledge—what machinery to buy, where to get lumber, how to operate the machines, and how to run an operation.

Using the $1,000 figure, Clyde convinced his family to invest in his idea. Start-up costs turned out to be closer to $5,000, but undaunted, Clyde approached Uncle Higbee, who didn't suffer spendthrifts gladly. Clyde's appeal must have been eloquent, for Uncle coughed up $3,000. James, his father, mortgaged the family home, and Clyde's mother must have gasped, but she never could refuse her youngest son. With his family's contributions, Little Brother was in business.

Clyde's brother Dette joined him in managing the family company. Brother Claude initially invested, but he was more interested in his sheep that covered the vast rolling hills west of Milton to the Columbia River. George and Darrell were orchard men, and Jim followed his father's penchant for easy money and gambling. But with Dette's wisdom, and his invested family members looking over Little Brother's shoulder, the company teetered off.

LOVE AND BOXES

CHAPTER 5

The company started with two bosses and one employee," Clyde would say. "Dette was manager and saw filer, and I was manager and operator. We had one employee, a 12-year-old kid we paid 10 cents an hour to run errands."*

The local gossip around town predicted that Milton Box Company would go broke within a year. But at the end of 1913 the little plant had grossed $13,000. Dette and Clyde paid themselves salaries for the year—$225 each. And they repaid their father so their mother wouldn't worry about a roof over her head.

Even though Clyde had gained a measure of knowledge about the box business from Al, the brothers were greenhorns when it came to building a factory—what equipment they needed, how to operate the machines, and, of course, where to get lumber.

The community largely stood by and watched, with one exception. Mr. Coyle, the city manager of Milton, also ran the power plant and was an electrician. It was his job to inspect the progress of this new business. Thankfully, he was an amiable man. Instead of criticizing their inexperience, he showed them how to wire the building properly and where to place the machines. At one point the brothers had diagrammed the flow of the lumber entirely backwards through the boxmaking process. Mr. Coyle pointed it out, instructing them on an efficient configuration. When the machines were delivered, Coyle even showed them how to start the motors.

They worked from one carload of lumber to the next, barely paying for the wood as the boxes sold. Clyde would remember the stress when the last of a carload of lumber slid through the saws, sanders, stitchers, then out, ready to sell as boxes. Where to get a steady supply of wood plagued their midsummer production.

"When we started in 1913, some people gave [us] three weeks to go

---

* Richard H. Utt, *Harris and the Pines* (Mountain View, Calif.: Pacific Press Pub. Assn., 1967).

broke, some three months, and few real liberal ones gave us six months," Clyde reflected years later.

A Walla Walla lumber broker had been quietly watching the fledgling business struggle for a toehold in the valley. He had seen and heard the hail of criticism raining on the young entrepreneurs. Why encourage a business sure to fail, some said. The place to make boxes is the Oregon coast, where sawmills are plentiful. One farmer stated loudly, "I will not buy from them [the Harris brothers] because the more I buy the more money they'll lose."

In the midst of the brothers' quandary on where and how to get the next carload, Mr. Robinson, a lumber broker, showed up. "Could you use a million board feet of lumber?" he asked.

The question was so unexpected that Clyde didn't answer for a moment and he didn't want to betray eagerness. Returning to the task at hand, he answered, "Maybe. But we couldn't pay for it until we sold the boxes."

"But could you use a million board feet of wood?"

"Well," Clyde replied, "We could use it over a period of time, but we couldn't pay until we sold the boxes. We couldn't even think about it."

"If we shipped you the wood, could you pay the freight on it?"

Clyde straightened up. "We could do that," he said. And the deal was struck with a handshake. It marked the turning point in Milton Box's struggle to gain its footing. As carload after carload of lumber lined up on the railroad siding beside the box factory, it took the entire Harris tribe two weeks to unload their bounty. Around the little factory towered beautiful stacks of lumber, a terrific advertisement to the community and an unmistakable signal that the Harris brothers were winning.

The stickler was the freight bill. For two brothers who had paid themselves only $225 each for a whole year's work, a $2,500 freight bill before profits looked almost insurmountable. But slowly the little box factory began to thrive and pay its debts. They conserved expenses by shutting down the operation in winter, Clyde and Dette finding jobs elsewhere. But the business slowly rocked its way toward success.

Twenty-three-year-old Clyde was also feeling his oats. Now that he was a full-fledged businessman, he bought a fashionable beige suit and a straw hat and had his portrait taken. And he was ready to fall in love. Dette was engaged to a girl named Marie, who served drinks in a saloon. Through the years she would become a family favorite with her easy laugh and good humor.

"That girl with the brown curls—the one in the middle," Clyde murmured one day to Dette with a tilt of his head as three girls strolled by.

"I want to meet her," Clyde added as he paused, hammer in hand.

"Well, Little Brother," Dette said, "she's pretty young. I think she's from the Frazier-Coe family. I doubt if you'll have much luck."

"I'll get my chance," Clyde declared. After all, he was a business partner in the box-making business—a detail he never let his family forget—and he knew that he was handsome. He conveniently liked to dismiss the fact it had taken the whole family to make "Little Brother's" business plan succeed.

Indeed, the odds were against 17-year-old Mary Viola Coe ever meeting Clyde Harris. Her father was a successful wheat farmer of 480 acres on Couse Creek south of Milton near the Walla Walla River. But Clyde was intrigued by the pretty schoolgirl with the confident air. She was soon to begin her first year of high school, the youngest daughter in a family of 10 children. Mary planned to become a teacher and was determined to let nothing detract from her education. She had plenty of friends and a high school beau named Roy.

Hers was a proud heritage. In 1868 her pioneer grandfather, William Samuel Frazier, platted and named their town Milton. He experimented with dryland wheat crops, wholly dependent on the natural wet and dry seasons, with great success. Mary's father would emulate his father-in-law's new farming methods to his advantage, too.

Twenty years before the Harris family stepped off one of the first westbound trains, Mary's mother, Susannah, had trudged west with a wagon train from Texas. William and Rachel Frazier would homestead 360 acres of land, and by the time the Harris family arrived, Grandfather Frazier had platted, chartered, and named the town Milton, donated land for a post office and a hotel, and had set Milton on a high tone as a temperance town. The Frazier farmstead would eventually become a museum and listed on the National Historic Registry.

When William Frazier's mules hauled the three wagons and family over the Rocky Mountains, his destination was the Willamette Valley. Their fifth child, Susannah, was 8 years old when her mother, Rachel, gave birth to her ninth child in a tent city on Nebraska's Platte River. Susannah never forgot the day the mule team that carried her mother and newborn Thomas spooked and ran wildly toward a precipice. Her father thundered in pursuit as the runaway wagon lurched toward the cliff, hauling up just sort of the chasm beyond.

The Willamette Valley beckoned, but when the mayor of Walla Walla met them at the Grande Ronde River marketing his fertile valley on the border of Oregon and Washington, they turned north to decide for themselves. Settling for the night under the stars on the state line between Oregon and Washington, they decided to stay.

Although both her grandparents were no longer living, 17-year-old Mary was accustomed to Grandfather Frazier's stern countenance staring down from the Coe parlor wall. His reputation for Christian kindness notwithstanding, he seemed to be frowning at her the evening that Roy helped her don her winter coat to step out with school chums for a little frivolity.

"But Mother, Roy and I intend to just watch, not dance," Mary sputtered in response to the reminder that dancing was a sin.

"Yes, Mrs. Coe," Roy added, pulling himself up to his not-yet-full height, the cowlick in front giving him an added inch. "Mary and I won't dance." And Roy kept his part of the bargain.

A newly opened second-story dance hall on the west side of Main Street drew the younger generation itching to welcome in the changing mores and times. It was 1914. Scott Joplin's smooth tones, including "The Entertainer," "The Ragtime Dance," and "Maple Leaf Rag," set the younger generation's toes of the tapping. The iconic Gibson Girls' tiny waists, frilly frocks, and curls escaping from prim updos were all the rage. And Grandfather's rules for a no-drinking town where frivolous pleasure was frowned upon was out of fashion. Milton had relaxed its rules.

With the devout promises, Susannah gave reluctant permission. What she didn't know was that a handsome, wrong-side-of-the-tracks young businessman was waiting his chance to meet her daughter.

Clyde was hammering out tunes on the piano in the corner when he looked up. Coming through the door on Roy's arm was the girl with the brown curls. Her eyes were bright, her breath was short puffs of frost, and her crimson cheeks matched her red mittens. His pulse quickened.

Suddenly it didn't matter to Clyde that a girl named Pansy was his date that evening. This was his chance.

Dette and Clyde had agreed to trade off on the piano, but when he signaled for Dette, there was no response. Three times he asked someone passing by as he played to do him a favor and invite "that girl" to the piano. If Mary got the messages, she didn't appear.

"Dette, take over for me," he hollered over the noisy crowd, nodding in Mary's direction. "There's someone I gotta meet." Dette finally took the hint and slid in beside his brother, seamlessly picking up the same chords and rhythm Clyde was playing.

With the polished air of a man who knew his way around, Clyde brushed past Roy and looked into Mary's eyes. "I'm Clyde. May I have this dance?"

Mary Coe blushed. Roy coughed. "We didn't come to dance," he said, trying to lower his voice an octave.

"Thank you," Mary said with a tiny smile. "But I just came to watch."

Her girlfriend nudged her playfully. "Oh, go ahead."

"Just one dance?" Clyde asked with a smile, his eyes never leaving hers.

"Well," she hesitated. "Maybe . . . but just one."

Clyde wasted no time. He whirled her into his arms and onto the dance floor. Years later Mary would still remember looking over Clyde's shoulder at her dejected beau.

With Mary in his arms Clyde led with an inspired flourish. Her heart raced. She was swept along with this handsome stranger, feeling the music, feeling the romance, the moment . . .

The music stopped. The magic passed. As he had promised, Clyde returned Mary to Roy. Embarrassed, she tried to hide her flushed face and compose herself. But it was too late. Her big brother Arthur stepped into the merriment just in time to see Clyde usher Mary off the dance floor. Mother Susannah had had an inkling of what might be going on and sent him to fetch Mary home.

And so the evening ended all too abruptly, but Mary never forgot those fleeting moments when her heart beat wildly after the young handsome fellow named Clyde. She never revealed what her punishment was. But the young man knew that if he intended to pursue this schoolgirl, he'd have to do it after classes in the daylight and not the dance halls.

Clyde began leaving work early every afternoon so he could be in front of the high school to greet Mary when school let out. At first Dette was amused. The hilarity and ribbing from his brothers intensified the day he bargained at

William Nichol's livery stable for a carriage and team of horses. He reined up in full view of the high school's front doors spilling with students and directly across from Milton Box Company. As the chivalrous romantic helped Mary Coe into his carriage for a ride home, the whole school was abuzz.

But Little Brother's neglect of duty left Dette shorthanded and was affecting the company's profit margin, so the family met to discuss his ardor and absences. They decided that Clyde wasn't tending to business. It was clear that if he had that much free time for romance during working hours, Milton Box Company didn't need Clyde Harris. The family board fired him. To add insult to injury, they delivered the news that he would have to buy his way back into the business.

With romance sidetracked, Clyde wheedled and bargained his way back into the Milton Box Company. He must have made some tall promises, and there is no record that he actually paid his way back into their good graces, but he returned to serious and steady hours making boxes. Still, after work, he made time to court Mary.

But Roy remained in the picture and made a gallant effort not to be outdone by his rival. On one memorable evening Roy called at the Coe home. The trouble was, Clyde showed up too. To cover their embarrassment, the two fellows tried to outdo each other conversationally. On the surface things were going well in the parlor. The hours ticked by. Mary thought about school the next day. But neither suitor wanted to be the first to depart.

It was midnight when Roy reluctantly caved in and stood up to take his leave. The next morning over breakfast, Mary's father had something to say about the parlor standoff.

"Well," he said, "I guess that Harris fellow finally won out last night."

Early one evening Clyde escorted Mary home from a baseball game, cupped her face in his hands, and placed a small kiss on her lips. "Will you marry me?" he asked.

"I'm going to be a teacher," she responded.

There were two more proposals, and each time Mary shook her brown curls no. She had plenty of time for future dreams. But when at last Clyde said, "This is the last time I'll be seeing you," Mary saw tears gather in his eyes.

"I wanted to make you my wife," he told her, "but I guess you have other plans."

"He got my attention," Mary remembered. "I told him I'd think it over and reconsider." The next time he asked, she said yes.

But Clyde's hurdles to win Mary weren't over. He dreaded asking her father and mother for her hand in marriage. The disparity between the life he had to offer Mary and her family's abundance was never far from his mind. When he finally got up his nerve, before the question left his mouth Mary's mother interrupted. "Why, Clyde Harris," she exclaimed. "Your mother never had enough food on the table!"

Perhaps Susannah had conveniently forgotten the modest uproar she had caused *her* parents, William and Rachel, when, at age 16, she fell for 31-year-old Samuel Keister Coe, a Civil War veteran from Northern Republic, who rode to town on a sleek black stallion. It was considered scandalous by Susannah's family that she should even entertain the notion of a Republican beau. Her father had served in the Confederate Texas cavalry, and the family's traditions were deeply rooted in the Southern tradition. Mother and Father Frazier must have swallowed hard before giving their consent to Sam Coe. Nevertheless, a big-city wedding with eight bridesmaids and eight groomsmen floated out onto the Fraziers' front lawn. One pioneer's wry remark, "It was almost all the young people in town," wasn't far off the mark.

Now with 10 children, the ninth one wanting to marry the young wrong-side-of-the-tracks hopeful, Clyde Harris, Susannah feared for her daughter, who wanted to marry a dance hall musician! Would his shoestring apple box business fail? If she married him, where would she live? Her fears were legitimate. But Clyde solemnly promised Susannah that Mary would always have plenty.

So it was that six months later, on July 8, 1913, that Clyde and Mary were married in a civil ceremony at the Walla Walla courthouse. Dette and Marie, who married a week later, had pressed for a double wedding, but Clyde didn't want to spend the money for a dark suit.

Clyde took Mary home to a tent he had set up beside his business. By tent standards it was magnificent. It was 9' x 12' with a wood floor and six-foot-high, double-sided sawdust-insulated walls. Clyde purchased a Franklin cookstove, a brass bed, a feather mattress, and a rocking chair for Mary. Mary slept under a feather quilt in the crook of Clyde's arm. At night winter temperatures dropped to single digits.

How did Mary survive? "I guess I was young and had good circulation," she would say. "And Clyde kept me warm in the feather bed at night."

RELIGION INTERRUPTS BLISS

CHAPTER 6

"I spent the first winter in the rocking chair with my feet propped on the oven door," Mary would recall in amusement. "We kept our canned fruit in a sawdust-filled basket in the corner to keep the glass jars from freezing."

One cold night well into their first winter, Mary briefly entertained the idea that she had married a madman. The couple was testing the limits of the durable wood stove and Clyde subconsciously worried about the flammable varnish he had applied to the tent walls as insulation. One night he dreamed their tent was ablaze. Still asleep, he leaped out of bed, grabbed the stove poker, and jabbed a hole in the tent wall. How he managed to drag his shrieking, unwitting bride out of bed and through the hole is a wonder. Before she fully comprehended what was going on, she was clutching at her flannel nightgown, bare feet trailing snow as Clyde hauled her wailing toward the creek.

The commotion roused Dette and Marie.

"Clyde, what in the world are you doing?" Dette shouted, shaking his brother. Groggy and confused, Clyde was suddenly awake. Had he saved Mary? Through the tear in the tent he could see the Franklin stove purring with a friendly warmth. Ruefully he took Mary in his arms. As they shivered in the snow, they decided it was a bad dream and began to laugh at Clyde's wild behavior. Soon they went back to bed, but it was quite a while before they could settle into slumber. Mary forgave her brave groom but didn't forget it.

It was customary for young couples to go to church on Sunday, and Mary began to anticipate the fun of walking into her family's Methodist gathering on the arm of her new husband. She didn't wait long before she announced, "We're going to church today." Clyde stirred in the warm bed as she primped.

"No, we're not," he mumbled and rolled over.

Mary glared at him. "Why not?"

"The only day I'll go to church is the Sabbath."

"Well, Sunday *is* the Sabbath."

"I mean the seventh day. I'll go to church on the seventh day."

She bristled. "You mean those strange folks on Advent Street?"

"I'll go to a church that worships on the seventh day."

"That's the Advent church. They hold meetings every Saturday. But it's not like regular folks who go to church on the day of the Resurrection."

"The seventh day is the Sabbath," said Clyde. "That's the only day I'll go to church."

"Well, shouldn't we at least talk it over?"

"The seventh day is the Sabbath," Clyde intoned, now sitting sleepily on the edge of the bed and smoothing his rumpled hair.

Mary's face reddened. "You tricked me," she yelled, stomping her foot.

They hadn't discussed religion before their marriage. She simply took it for granted they would join a respectable Protestant church. What an embarrassment! Her handsome, worldly-wise groom had become a stubborn religious fanatic.

"If I'd a' known that, Clyde Harris," she shouted, "I'd 'a never married you."

"The seventh day is the Sabbath," he intoned again, and Mary's blood boiled.

"I guess I sassed him," she recalled with a wry smile 75 years later. "But he made me so mad. He just kept repeating over and over, 'The seventh day is the Sabbath.' "

She stomped, sassed, cried a little, and went home to tell her mama how awful Clyde was to her. Mama Susannah, a woman of the early twentieth century, told her ninth child that now that she was married, she should listen to her husband and go to the church he chose.

Mary might have been a Victorian Age bride who had learned a woman's place in those times, but that didn't mean she parked her opinions, her religion, and her spunk at the courthouse steps when she said, "I do." Clyde was acting righteous. Well, she would outdo him on that score! Why not invite a pastor of a Sundaykeeping church for a chicken dinner each week. She was sure the reverends' discussions would set her misguided, stubborn husband straight.

As Mary's plan unfolded week by week, it must have been with some relief that Clyde caught wind of a series of meetings at the Adventist church. He made sure he and his protesting bride attended, which, given Mary's diligent religious entertaining, meant their weekly evenings were crowded. She was protesting, all right. Clyde insisted his brothers and sisters attend the evangelistic meetings, too—for their mother's sake, he argued. His oldest sister, Martha, must have been amused and grateful for Little Brother's success. She had joined her mother's church long ago, but her younger siblings had always been uninterested. That is, until Clyde persuaded them.

But Mary didn't give up easily. She was determined to prove her husband wrong.

With Clyde teaching her how to cook, her fried chicken was mouth-watering, much to her husband's pride. In addition, the soft Southern charm she inherited from her mother made her a consummate host.

As the weeks rolled by, she entertained one after another of the local Protestant pastors. They came for an evening of good food and a discussion aimed at Clyde as to why he should change his rigid mind about his day of worship. As the theories of Sundaykeeping abounded, Mary realized that though the preachers were gracious, her husband remained unmoved and unconvinced.

Finally, her flock of chickens depleted and no more fowls to fry, but still exasperated, Mary invited the Adventist pastor for a meal. "It won't be a chicken dinner," she told him with a touch of petulance. "I've used up my flock on the other ministers in town."

The pastor laughed and said, "That's just fine with me, Mrs. Harris."

By the end of the evening Mary was charmed by the minister from Advent Street. She also appreciated his sincerity and his knowledge of the Bible. So in 1915 Clyde and Mary, along with Dette and Marie, Claude, Darrayl, Maude, and Hazel, were baptized. Mother Bertha was a happy woman.

The small group of Sabbathkeepers felt it their duty to teach these frivolous new converts just how serious religion should be. As stouthearted pioneers who had braved the trip west with large families, *they* knew the rigors of life, and tact wasn't a plentiful Christian grace in those tough-minded times. The early Northwest Adventists were independent thinkers.

Mary wasn't sure she would like it. On her first Saturday in attendance she was resplendent in a new green felt hat she had made herself, pheasant feather swooping over her face. She felt a certain pride that she was brightening up the drab congregation. Sitting respectfully in the pew, Mary felt a tap on her shoulder as a church sister slid in beside her. Her head swiveled, the pheasant Feather brushing Mrs. Standard Bearer's pointed nose.

"Honey," the woman crooned, "we don't wear outward adornment at this church."

Mary wasn't amused. But in her ninetieth year she enjoyed telling the story. She lost the feather, kept her temper in check, and grew to love the stouthearted saints who met and worshipped on the seventh day of the week. She came to realize the hardheaded pioneers needed grace just as much as she did. After all, her grandfather Frazier had been a stalwart pioneer who had raised the ire of not a few settlers when he declared Milton a dry town.

Mary, the gracious new convert, gathered children in her arms, taught their Sabbath school classes, and sang songs to the toddlers. She always loved it. Clyde didn't get a pass either. The nominating committee added his name to the list of deacons. He nearly perished with fright the Sabbath he had to give his first morning prayer. When they appointed him a Sabbath school teacher, he quickly became a diligent Bible student, fearfully facing the rows of oldsters sitting before him. The church leaders later admitted they put the elder members in his class by design. "It was their effort to teach me," Clyde remembered.

Their joys of fellowship grew over the years. Orchardist Aaron Miller's sons were among their favorite friends. Never in her wildest dreams did Bertha imagine that her firm convictions would lead to such friendships among her children, reared as they were on the other side of the tracks. In the little white church that Bertha had joined so many years before, everyone was equal as they contemplated the Savior.

BOXES AND BURDENS

CHAPTER 7

The factory was gaining in orders and profits. The Milton *Eagle* announced: "Last season the local plant used about a million and a half feet of lumber, and expects to use fully a third more than this during the season of 1916. They have purchased a sawmill at Wahkiacus [on the Columbia River]. Work will commence on an order for two carloads of lettuce boxes for the greenhouse at Walla Walla River."[1]

By mutual agreement the two couples would spend one year each at their new sawmill on the Columbia River, trading places at year's end. Dette and Marie were first to move there, because Mary was pregnant. As the couple rejoiced in blissful expectation, Dette and Clyde decided to consolidate operations and build a sawmill at Milton. They sold the Columbia River plant.

Mary wanted many children. Her mother, known with great fondness in the Milton Valley, had set a cheerful example of having 10 children. When she died at age 89, the Milton *Eagle* extolled the pioneer grandmother saying, "Mrs. Coe, known to most people as Aunt Susie, met everyone with a welcoming smile. . . . The more crowded her home, the happier she was."[2]

Although pregnancy came with a high risk in those days, the big families of both Mary and Clyde eased any cause for worry. But then everything changed. Suddenly Mary began having excruciating abdominal pains, then started to hemorrhage. Something was very wrong. At the hospital an emergency surgery saved Mary's life, but not the life of her baby.

The diagnosis was an ectopic pregnancy. On rare occasions the fertilized egg fails to lodge properly in the uterus and instead attaches to other tissue outside the womb—in Mary's case, a fallopian tube. A ruptured tube, a mistake of nature, was often fatal in those days.

[1] Milton *Eagle*, Feb. 25, 1916.

[2] "Pioneer Milton Woman Passes," Milton *Eagle*, Jan. 2, 1947.

Clyde faced the frightening experience with husbandly terror. The doctors assured the couple it was a rare occurrence and not likely to happen again. Clyde took Mary home to recover.

But it *did* happen again. Her second pregnancy went well—until her third month. Then came the blinding pain, the terror, the emergency surgery. Mary narrowly survived. To a distraught Clyde the idea of her conceiving again was out of the question. He simply would not risk her life again. Mary might have been less certain, given the finality of it, but Clyde and the doctors made the irrevocable decision while her life hung in the balance. The decision took the best of her dreams. Clyde's vision for the future was of expanding the company, but Mary had centered her life on her home and having a big family. It was a bitter pill. For the rest of her days her face bore a wistful longing for what was not to be.

The tragedy of Mary's childlessness was the second heartbreak in her young life. Mary was only 10 years old when she lost her oldest sister, Berniesa. On a cold December night while the family attended prayer meeting, 17-year-old Berniesa stayed at home with her brother, 8-year-old Harold, saying she needed to study. The newspaper account tells the story.

"One of the saddest affairs that ever occurred in Milton was the tragic death of Miss Berniesa Coe, who died last Sunday evening from the effects of carbolic acid administered by herself. It is believed the awful deed was caused by extreme nervous depression caused by overstudy and mental worry.

"Miss Coe had been doing extra work for several weeks, sitting up until 12:00 and 1:00 every night preparing for the examinations at Columbia College, where she was attending school. Sunday evening she remained at home to study while the rest of the family, with the exception of a younger brother, went to church. About 8:00 she arose from her seat and, going into another room, soon returned with a small vial and drank the contents. She then lay down on the lounge with the remark 'I am going to heaven.' Her brother, seeing that something was wrong, summoned help. Her parents at the Methodist church were notified and Dr. Thomas was summoned, who in turn called in Dr. Dotson to assist in administering an antidote, but nothing could counteract the effects of the poison. The patient died a few minutes after 11:00, having been unconscious from a few minutes after taking the poison until the end.

"Funeral services were held in the Methodist church. . . . Columbia

College was closed and the students attended in a body. . . . The casket was attended by 10 young ladies in white. . . . Business houses were closed and the large auditorium of the church was unable to hold those who attended to pay the last tribute of love and respect to the deceased. . . .

"Miss Coe was one of Milton's most lovely and beloved young ladies and her tragic death has cast a pall of gloom over the community."

Carbolic acid, the newspaper account read. But why? The grief-stricken family was frozen in disbelief. Berniesa, who had cuddled little sister Mary every night in the bed they shared. How could she so unaccountably take her own life while her parents attended prayer meeting? The whispers started. Perhaps she was pregnant, and hopeless, they inferred. The family chose silence to protect Berniesa in the best way they could, but the tragic circumstances made it difficult for the family to grieve in the normal way. They merely stopped referring to her. Unless one looks very closely, she is simply omitted from the family records.

After Mary had lost two babies, Clyde became her world. She would be forever frozen in a girlishness that only thinly veiled the sad losses she endured at such a young age. Clyde did his best to make it up to her as she mourned in what must have been a prolonged postpartum depression. He tried to lighten her mood by suggesting, as Elkhanah of the Bible said to his barren first wife, "Aren't I better to you than 10 sons?" He bought her a cute little white house in the center of Milton just a few doors from Dette and Marie. With Milton Box making more profits each year, he could afford to indulge his bereft wife, and she restored and decorated it to her heart's content.

Gradually, as she recovered, she discovered she loved the renovating process and reveled in her newfound abilities to design and build. She found she was good at it. When they moved to Pendleton, she designed the almost 80 houses in Clyde's new subdivision named Harris Heights.

As Milton Box Company gained success, the naysayers had to admit the local boys had made good. Even after the hot summer day in 1924 when the box factory caught a spark and burned to the ground, the partners secured financing and quickly rebuilt. The new factory transformed the business into a sturdier, more modern company. Nothing, it seemed, stood in the way of their progress.

Then in 1925 Clyde embarked on an ambitious business venture. A local contractor was bidding to build a Catholic school in Walla Walla. It was to be a lucrative job. If Milton Box Company provided the lumber, they'd make a handsome profit. The contractor submitted the lowest bid and was awarded the job. Clyde and the contractor agreed on terms. But there was a big problem. Midway into construction the contractor ran out of money and couldn't pay the box company. Too late the contractor realized he had failed to add the roof into his bid price.

The contract was inviolable. To renege would bring the company the scorn of the community. Business leaders would view Milton Box as unreliable. With dismay, Clyde realized he had to absorb the monetary loss with his personal assets. He simply couldn't finish the contract, including the school roof, unless he sold their newly renovated house in the town center. Clyde broke the news to Mary and drove a hard bargain with the contractor. He would exact the contractor's house on rural Winesap Road as his part of the deal. Mary had to move.

As a consolation for them living so far from town, he bought Mary a car. And although at the time it seemed like a great trial, the move would prove providential when the Great Depression hit. Under Clyde's tutelage, Mary raised acres of tomatoes and flowers of all kinds. Her fruit and flower stand kept food on the table during the lean years that soon were upon them.

But the mishap's collateral damage was a toll on the couple's relationship. Mary would ever be hypervigilant regarding Clyde's spending without her knowledge. "Don't tell Mary" became a phrase he used often as he made deals and at his occasional generous impulses. Regardless of his wife's fears, he simply couldn't quench his gambling spirit in business or, at times, his generosity.

FACING INTO THE STORM

CHAPTER 8

By the end of the 1920s Milton Box Company had weathered many adversities and was grossing more than $300,000 a year. Two shifts and nearly 100 employees kept the company rolling as it rebuilt from the 1924 fire. It wasn't hard to believe good times were here to stay.

The Great Depression wasn't supposed to happen. New wealth had lulled many into complacency. But in 1929 the Gilded Age lost its luster. Everything changed. The lights on Broadway went dark, chorus girls no longer kicking high in musicals. The stock market collapsed, and overnight multimillionaires became paupers. The domino effect created a worldwide economic meltdown.

The collapse oozed west like a thick sludge, quelling the heartiest businesses. Jobs were cut. Banks were forced to declare insolvency. Wild progress slowed to a crawl. Workers lost their jobs. The promises of good years here to stay dissolved as more and more naysayers predicted a permanent shift in America.

Clyde and Dette faced the wrenching reality as Milton Box struggled to survive. Their bank in Portland, the stalwart backbone of early financing days, demanded immediate repayment of a $10,000 five-year loan. The brothers decided on three emergency measures:

1. They poured all their personal money and securities into the business to keep from closing the doors.
2. They slightly reduced employee wages.
3. They issued scrip for 40 percent of paychecks, redeemable at local merchants at a 5 percent discount or, if held six months or more, payable in full value.

Of course, the second and third measures weren't popular with the wage earners, who struggled to feed their families. Some grumbled. Their paychecks gave them only 60 percent cash and 40 percent in redeemable scrip. They complained that the company was raking in profits off its employees, who had no choice but to accept the discounted scrip in these hard times if they wanted a job.

But others deemed it a brilliant move on Clyde's part. The company stayed in business, and the employees kept their jobs. If they didn't like the discounting, well, they could try to find other work. One man found a sawmill job in La Grande, left the Adventist Church, and for a number of years bitterly regretted his decision. Years later Clyde hired him back and promoted him to foreman of the furniture factory. The grateful fellow was rebaptized in the Pendleton church.

But as the Depression pressed mercilessly upon America, the brothers hunkered down in survival mode. In 1932 the company wrote off $75,000 in bad debts. That year the company losses were $6,000. Ever after, Clyde lamented writing off the loss that sullied the company's perfect record of profits.

He never forgot how his hand shook when he opened the day's mail. His knees would buckle when he tore open a creditor's envelope and read the bold words "Immediate Payment Demanded."

"When I received a telegram, I'd have to sit down to open it. I was just too weak to stand up," he would recall. "I'd say to myself, 'More bad news, huh?' and shake like I had the palsy, wondering what could be in it."

"I tell you this," he added, "to give you a little idea of what a man goes through in those kind of times when you have one disappointment after another so often that you begin to wonder what's next." Then in reflection, he said, "But the Lord took care of us. He gave us health, which we are very thankful for."[1]

"In financing our business from 1932 to 1934, we owed the banks and financing houses a lot of money, and we couldn't pay them. I didn't know whether I could stand it or not. I went to bed at night and knelt down to say my prayers; I wondered if that would be the last time I would be saying or knowing anything. But," he continued, "I told the Lord, 'You have led us along this far. If You want it to succeed or fail, I will accept it. I turn it all over to You.' Then sleep would come."

It was in those crisis years that Clyde's faith deepened. "I always followed the principle 'If any of you lack wisdom, let him ask of God, that giveth to all men liberally.' That's in the book of James. [God] has always given this wisdom. Every time we came to a blank wall, a way would open up."[2]

---

[1] Milton *Eagle*, c. Jan. 28, 1908.

[2] R. H. Utt, *Harris and the Pines.*

The couple faced hardship together. There would be no salary during the Great Depression, but they again drew close as they prayed through the dark years. Clyde resolved he would always "do right by the Lord." He put his gambling spirit aside and found strength as he listened to the Holy Spirit.

THE BLALOCK STORY

CHAPTER 9

Walla Walla College, Clyde Harris, and the young toughs of early Milton had Dr. Nelson Blalock to thank. In 1873, while Mary's grandfather William Frazier was platting, naming, and chartering Milton, the Illinois resident and Civil War surgeon arrived in Walla Walla for a fresh start. His life had not been easy. He was haunted by memories of Civil War battlefields, his near death from typhoid and dysentery, and the deaths of his wife and eldest child. In the vast wilderness of the West, he, as the only doctor, practiced itinerate medicine, sometimes galloping 150 miles on emergency calls.

But it was his fascination with improving the area that made him a familiar name in the Walla Walla Valley. Dr. Blalock turned orchardist, planting on land considered too arid for growing anything. He dug wells, planted apple and pear orchards, and fruit stock from Aaron Miller's new nursery. He produced bountiful crops and, following William Frazier's example, put his hand to dryland wheat farming.

He served two terms as mayor of Walla Walla, during which time he listened with interest to the Adventist Church leaders' pleas to build a college in the area. Then Blalock extended 40 acres of his own fruit orchards to the Adventists free and clear.

As the population of the area increased, he also noted the need for building materials. Lumber was essential for building houses, barns, and fences, but it was with much difficulty that timber was transported long distances in horse-drawn wagons.

A steep, white fir-covered mountain, elevation 4,578 feet, and a 13-mile buggy ride from Milton, was Blalock's answer to obtaining timber within a practical range. Piece by piece he hauled an old sawmill to the summit. To move the fallen trees down the mountain, he embarked on a massive project, building a water flume, eight feet wide by four feet deep, that snaked down the curvature of the mountain for 28 miles. It cost

$56,000 to build. Logs, lumber, and fence posts floated down the flume to the North Fork of the Walla Walla River, then were horse-drawn into town. The huge operation was the talk of the Northwest. At the top of the mountain Blalock installed the first telephone in the state of Oregon to connect to the flume end in Milton.

Unfortunately, when the project was completed, he was $150,000 in debt. So in 1882 he sold the mountain and the industry to the Oregon Improvement Company. Loggers and their families continued to live on the mountain in a tent city. Their children attended a small school on the mountaintop.

In a reflection on the history of Milton-Freewater, an old-timer reported, "Rode Edwards recalls riding lumber down the O. I. Company flume on lumber and cord wood with Joe Dykes, Det Harris, . . . and other young sprouts of the village."[1]

Of course, jumping astride free-floating logs was not allowed, but boys have a way of getting around rules. Perhaps the log riding tales explain the fascination the Harris brothers had with the pristine mountain that rose high and lofty almost in their backyard. Whether Clyde took the daredevil ride or not, it's a sure bet that he heard Dette and the boys talk and his imagination took the wild ride. And one day he would build a modern "wild ride" down that same mountain.

It's little wonder that by the mid-1930s Dette and Clyde would turn deals until they became the owners of the mountain now called Blalock by a series of land buys, exchanges within the family, and, as one old-timer asserts, "a big win in a poker game," although that is entirely anecdotal. But the truth is that Milton Box, soon to become Harris Pine Mills, gained a mountain of spectacular timber, which led Clyde to remark to Charlie Nagele in 1951, "As long as you have Blalock, there will always be Harris Pine Mills."

Nelson Blalock's foresight paved the way for the logging road built by Milton Box Company in 1937. No longer dependent on horses and mules, logging was engine powered. The road was a wonder to the community.

"'One foot in the grave and the other on a banana peel, but I like it,'

[1]"Random Collections," in *Early History of the Milton-Freewater Area* (originally published in *The Valley Herald*, 1962), p. 25, http://museum.bmi.net/EARLY.HISTORY.M-F.AREA.htm.

commented the man who piloted a mountain of logs on a big truck down 3,000 feet from the top of Blalock Mountain to the Milton Box company sawmill. From where the logging road leaves the South Fork canyon, three miles above the powerhouse, it is six miles to the logging operations on top. A large part of the way the grade is 20 percent, some short stretches 25 percent. . . .

"At the summit you enter a dense forest for about two miles over a good dirt road at the end of which you come to a tent camp of half a dozen loggers. Blalock now presents a thicket of second growth, among the scattering larger trees, resulting from the early sawmill operations of Dr. N. G. Blalock. A huge Caterpillar wades through the pole thicket pushing over trees a foot in diameter, hitches onto a half dozen saw logs and drags them out to the loading derrick.

"With a permit from my old friends Det and Clyde Harris of Milton Box company, I took a ride up Blalock on one of the super colossal logging trucks, the latest devices of thundering mechanical power, making possible the seemingly impossible. . . .

"It really sounded fantastical when the Harrises announced they were going to build a logging road up Blalock, and even after the road was completed with the steep grades, it was commonly remarked that logging down this road would be impossible. However, since the road became a reality six months ago, three trucks with 16 big tires, and water-cooled air brakes, have been making two trips daily, hauling 4,000 board feet of logs. With the seeming great risk, these dare-devil drivers possibly take little more chance of casualty than a 70-mile-per-hour car or a pedestrian crossing the street."[2]

One day Mary would need furniture for the cottage they built atop the mountain. And the door would open for another giant step for Harris Pine Mills. It was impossible for Clyde to think small.

---

[2] L. W. Roberts, "Logging of Blalock."

READY FOR PRIME TIME

CHAPTER 10

By 1935 the company had weathered the Depression and was growing again. Maybe it was presumptuous, but given the other details of Clyde Harris's life, it was more like divine serendipity that the two brothers entertained the notion of bidding for government timber that was all but sewn up by the well-heeled Mount Emily Lumber Company just across the Blue Mountains.

With the demand for lumber climbing and World War II looming ahead, the U.S. government began offering their prime timber as a tool for community development. In 1937 Mount Emily Lumber Company, located in La Grande, Oregon, held the upper hand bidding for government land in the forests of the Blue Mountains. Years earlier Wisconsin lumber barons Stange and Kinzel, finding the available Western timber plentiful and cheap, had pushed west to establish the lumber company. Other Midwest lumbermen were bidding and buying northwest forests as the timber in Middle America dried up.

For a number of years Mount Emily Lumber enjoyed no competition. They won bids from the Umatilla forestry division and built railroad lines into their property. Their good luck led to overconfidence. With two recent bids won from the Umatilla forestry division, the company built railroad lines into their property. In anticipation that they would successfully win the Camas Creek block, the LaGrande timber barons forged their rail lines into the land coming up for bid. Then the prosperous lumber company approached the forest service requesting "the remaining National Forest saw timber within the Camas Creek Block of the Umatilla National Forest."[1]

The result of this quiet request was duly authorized by the Washington, D.C., Office of the Forest Service, and the prospectus for the timber sale was issued. The city of La Grande was grateful for the economic

[1] David C. Powell, "The Camas Creek Timber Sale and the Milton Box Company," June 2008, https://fs.usda.gov/Internet/FSE_DOCUMENTS/fsbdev7_015648.pdf.

stimulus Mount Emily Lumber provided. So with the wind at their back, the lumber barons of Wisconsin believed the sealed auction was in the bag.

The Forest Service approved of their high hopes for another easily won bid, but the company decided to hedge its bets. The prospectus that had been so carefully surveyed by a large group of eager forestry students was only the legal formality. Mount Emily wanted to ensure they didn't bid too high. After all, even a low bid would get them the whole 221 million board feet of ponderosa pine, winning a no-contest submission—that is, if no other company was interested.

At a meeting in Burns, Oregon, on June 3, 1937, officials from the Mount Emily company met the Washington, D.C., government officials. Both sides assumed that since the last two bids on the government-offered land were unopposed by other bidders, the sale would be a slam dunk for Mount Emily. However, the wording of the prospectus, as required by federal law, stated it was open to bidders. The prospectus from the government stipulated that no bid under $2 per thousand board feet would be recognized for the prime timber being offered.

Just over the Blue Mountains another community, Milton, needed timber too. But Mount Emily, with its deep pockets, casually overlooked the little box manufacturing business in the Walla Walla Valley. Well, almost. They wanted to be sure.

One morning a man wearing a white Stetson and assuming the air of a well-heeled landowner appeared at Clyde's office. He began by saying, "I own 140 acres contiguous to the Camas Creek government auction coming up."[2]

Casually, though he didn't go so far as to pick his teeth with a blade of straw, he displayed the disinterest of a farmer merely throwing out an option for Clyde and Dette to ponder. "I thought, since you would be participating in the bidding process, that you might be interested in buying my land as an advantage to you if you were to win the bid."

Clyde's antennae went up. Something about the guy didn't square. He countered the curious stranger's pose with the disinterested demeanor of a poker player with a winning hand. Whoever this man was, he wasn't schooled in the ways of a gambler.

"I think you've come to the wrong company," Clyde told him. "Mount

[2] R. H. Utt, *Harris and the Pines.*

Emily is the company who will most likely win the bid and would be interested in your proposal."

Whatever the undercover operative took back to his bosses at Mount Emily, he undoubtedly conveyed his impression that Clyde Harris didn't exhibit any interest in the Camas Creek government offering. If Clyde and Dette had been thinking of entering a bid, this latest turn of events and the Harris brothers' gambling savvy spurred them on. They took the plunge.

In view of the government prospectus notifying all relevant parties of the proposed sale, the 19-page document showed a clear bias toward Mount Emily as the only viable bidder and listed a long string of reasons they should be the logical winners. It discussed the fact that the Forest Service wanted to set a minimum bid with Mount Emily's profit line as a clear factor. But the government could not see any bid on the primary timber, ponderosa pine, going for anything lower than the minimum bid of $2 per thousand board feet. That was the stickler. Mount Emily intended to cast the minimum bid, sure of no opposition. The not-so-covert scout who came probing Milton Box's intentions reported to his company that the bid would be entirely theirs.

Perhaps because the scout might recognize Clyde, they agreed that Dette would deliver a sealed bid in person at the designated Portland, Oregon, postal site. He ensconced himself in a hotel near the post office and made himself scarce until the 2:00 p.m. deadline.

At 1:58 p.m. Dette Harris stepped into the post office and dropped his sealed bid and $4,000 cash on the table. The bidding closed, and a small crowd gathered their wits and waited. Government officials duly authorized and, with unprejudiced demeanor, opened the bids.

Mount Emily's bid was read first. Two dollars per thousand board feet, the minimum government price. Then Milton Box's bid was opened. It was $2.16 per thousand board feet. The place began to buzz. People shook their heads. This could not be happening. Milton Box Company had just won the government offering of Camas Creek and almost 50 square miles of timber.

A ruckus ensued. Word flew to La Grande. Incensed timber bosses, community leaders, and citizens alike joined in the chorus of "Unfair! Unfair!" They petitioned that the process be reconsidered. Mount Emily had built rail lines into the government-forested timber and was in the process of establishing a rail into Camas itself. They whined and argued that they were supposed to be the recipients of the government's largesse. They sued. They appealed to the U.S. Department of Agriculture. It just wasn't fair that this little box company would win the bid.

PENDLETON PERSUADES

CHAPTER 11

With uncommon foresight Clyde and Dette had stealthily laid groundwork should they win the Camus bid. Indeed, Milton-Freewater rested overconfidently on the fact that their homegrown boys would never leave the comfortable environs of the valley.

Now a problem developed. Two campaigning city council candidates proclaimed that, if elected, they would see to it that Milton Box Company paid more for power, and Clyde and Dette took it personally. The Harris family had largely ignored the town's class-consciousness. With their talents and energy, the siblings had launched a variety of lucrative pursuits besides boxmaking—orchards, sheep, wheat, and, on occasion, a successful gambling spree. Milton Box Company was now timber-rich with the logging rights at Camas Creek. But they needed to be on a direct east–west rail line. With the alignment of their prosperity, the possibilities of change outweighed the secure moneymaking setting where they had started it all.

Rumblings about Milton Box Company's local unease reached Ed Aldrich, editor of Pendleton, Oregon's *East Oregonian* newspaper. The long-revered newsman was a leading voice in Pendleton's city council, the civic clearinghouse in the town 40 miles south of Milton.

Reading the comments in the weekly Milton *Eagle,* Aldrich sized up Milton's political climate. Then he went to work gathering pledges and resources to make an attractive offer to the Harris brothers. With the underlying principle that would become the grand-opening slogan for the relocated company in 1940, "What's Made in Pendleton, Makes Pendleton," the city opened their arms to Milton Box—that is, if a gift of 36 acres and $10,000 felt about right to the Harris brothers. It didn't take much wheedling to get Clyde and even Dette to listen.

With the deal well defined and Dette in agreement about the move, Clyde brought the news to Mary. "The city is offering us 36 acres of land and $10,000 cash to relocate our plant," he told her. Her lack of resistance surprised him. Maybe Mary was restless at the Winesap Road home she had been forced to accept upon Clyde's unfortunate collaboration with the high school construction project in Walla Walla. The Pendleton offer coincided with their twenty-fifth wedding anniversary and excited Mary with the future possibilities. The possibilities for growth were exciting. Still, the decision lay heavily on Clyde's mind.

In an undated "hunt-and-peck" typed memo, Clyde wrote a barbed summary that brought them to the decision to move their company to Pendleton. He reviewed the relationship between Milton Box and its town.

From the Desk of Clyde H. Harris.
Review of our operation:

When we started in 1913, some people gave three weeks to go broke, some three months, and a few real liberal ones gave us six months.

About one year later one man came to me and said, "Clyde, is it possible that you can buy lumber and cut it up and make a profit?" He still could not believe we could.

In the thirties two councilmen went in office in Milton on a campaign of making Harris pay more for power if they went in.

About two years later these same men offered to cut our rate in half if we would stay in Milton, and the rate had not been raised either.

When we refused, they asked what we would take to stay. We had made a commitment to Ed Aldrich to move to Pendleton and could not change.

We told Mr. Aldrich:

- We close on Saturday.
- We run our plant on Sunday.
- There will be some smoke from our plant.
- Our people do not believe in unions.

Ed answered, "We have taken all of that into consideration in our meetings, Clyde, and we want you to move to Pendleton."

I said, "Ed, if we can't come now, we might come later."

He answered, "We want you to come now."

Real men make cities; they do not tear them down. We came because of Ed Aldrich.[1]

---

[1] Clyde H. Harris, personal notes, 1940, in author's possession.

More and more, the vision for the company's future fell to Clyde. After weathering the Great Depression and bearing the excitement and trepidation of the Camas Creek buy, Dette was feeling all of his 60 years. Twelve years older than Little Brother, he wanted to back away from the unrelenting highs and lows of building the company. It was increasingly difficult for him to subordinate his thoughts for the company to Clyde. He groused to a few family members that when he had a good idea, Little Brother would ignore him and later hatch it as his own plan. In addition, although family loyalties would always remain strong, Dette was finding it hard to move at Clyde's lightning speed.

Before the deal was inked, Clyde was already worrying about the critical housing shortage in Pendleton. A company move meant approximately 50 families would need lodging. Margaret, Mary's niece, remembers that "in 1939 when the mill was moved to Pendleton, the country was poised on going to war. . . . With the Depression and now a war looming, it had been 10 years since any housing had been built in Pendleton. Addressing the practical problem of relocating their operation to Pendleton, Clyde bought acreage on South Hill to build a subdivision of small but adequate bungalows for the influx of workers."

As Clyde revealed the impending move to Mary, he hurried to tell her about the hill on the south side of Pendleton. He had walked the whole of the property for sale, and the price was right. He explained how he planned to build dozens of tidy homes that people who worked for them could buy at nominal prices. And Mary was to design each house. Elmo, her brother, would be the supervisor and electrician. No wonder Dette wanted to get off the bullet train.

Mary couldn't resist Clyde's logic, his zest for the latest plunge, and its possibilities. Building would begin almost immediately. Mary embraced the project, becoming a blueprint designer. Every lot elevation presented a different problem. Bending over the kitchen table night after night, she drew the plans, figured the topography, and put function into form as Clyde divided the land into fiscally feasible lot sizes. Mary was proving herself as her husband's right hand as she designed these bungalows in record time.

In 1944 Elsie Dickson wrote about Mary's accomplishments in the Milton *Eagle*:

"'It wasn't so much indulging a hobby as meeting a necessity,' said Mrs. Clyde Harris when asked how she happened to draw the plans for 50 houses, mostly in Harris Heights, built by the Harris Pine Mills, of which her husband is president.

"She said that when the Heights project began in 1939 (the project of 62 houses was completed in 1942), there was a tremendous job to do. Pine Mills employees had to have a place to live, and in some cases as many as six houses were under construction at one time.

"Though she has had no training in the field of architecture, Mrs. Harris said that ever since she was a child she has loved to draw house plans. Born in Milton of pioneer parents (her mother, Mrs. Susan Henrietta Coe, 87, of Milton, crossed the Plains in 1877), Mrs. Harris, after she and Mr. Harris were married, lived in Milton, where she drew plans for two homes that they remodeled.

" 'That was my first practical experience. When I tackled the job here, I was in for some hard labor, and I'm glad it's over—so I can enjoy myself,' said Mrs. Harris."[2]

As Clyde moved out on Aldrich's proposal, Pendleton was delighted with the injection of spirit, business opportunities, and jobs that the Harris brothers brought with them. Clyde and Dette burned up the road between Milton and Pendleton as building began in earnest in 1939, first a box factory, then a sawmill. Heavy equipment carved a log pond, and a mechanic's shop and office were quickly thrown up. Mary hurriedly sketched out plans for their first house on the corner of Ninth and Byers streets, took a little time out in August for their silver wedding celebration, and by Christmas Eve they moved into their new house.

Word was out. The rowdy town 40 miles to the southwest was about to expand. Kip Cady of the *East Oregonian* would remind Pendleton just how it happened:

"Santa Claus was shining his boots about the time the man stepped up to the saw. On Christmas Eve in 1939 the impending arrival of the Jolly Old Man was about the only cheery news residents of Pendleton and the rest of the world had to look forward to. War was brewing in Europe and the Depression still nagged the economy.

"The man standing at the saw had things other on his mind. He had, during the year, moved his box manufacturing plant from Milton (now part of Milton-Freewater) to a 36-acre site in south Pendleton at the behest of the city's community leaders.

[2] Elsie Dickson, Random Reflections, Milton *Eagle*, Mar. 18, 1944.

"The clock said 5:30 p.m. when the saw was initiated to its first log cutting—some spirited worker had even inscribed the tree 'Merry Christmas.' . . .

"Harris, himself, cut the first log that Christmas Eve Sunday in 1939. Fifty workers were employed at the beginning. By May 1940 two hundred were on the assembly line."[3]

Clyde braved the sleet and shouted instructions as his men wrapped chains around the first giant ponderosa and winched it up into the tin-roofed sawmill. Taking the thick plank stairs three at a time, he raced up and jumped into the seat of the carriage rig to personally saw the first log.

Steam-powered dogs, steel jaws, bit into the log and held it in place. The band saw, all 16 feet of it, sliced an eight-inch-wide swath off the massive log, cutting through the limbless trunk as if it were butter. With an awesome rumble the slab flopped onto moving steel rollers. Exhilaration filled the air with the scream of the machines, the pungent pine and fir sawdust raining down. The rig and rider flew back and forth. The noise heralded the news. Harris had come to Pendleton!

A new history began that would change Pendleton and infuse new life into the tiny local church. And one day business circles nationwide would fondly refer to the company as "the church." This business would one day dedicate all its increase to Christian education and other noble causes around the world.

The machinery worked seamlessly. The man who envisioned this company could not have had a better Christmas present. Meanwhile Mary enlisted the help of her niece Margaret and Margaret's new husband, John, who was bound for medical school the next year. The three of them lugged boxes and furniture between Milton and Pendleton. But Clyde neither trimmed a tree nor kissed Mary under the mistletoe. "Why let the holidays interfere with business?" was Clyde's reply to any questions about how he planned to celebrate.

Milton Box Company was changing its name. The apple box business was emerging as a broader wood industry. More and more the company was becoming Clyde's baby as, one by one, his brothers stepped away from

[3] Kip Cady, "Harris Pine Grows From Box Plant," *East Oregonian*, June 30, 1980, p. 15P.

their advisory positions. Dette stayed on two more years until his ill health demanded he retire.

With the worker migration the vise tightened on the strapped housing market. Stopgap housing was needed immediately. In his practical style Clyde co-opted the southern side of the mill to throw up logger shacks in record time. Each had enough room for a cookstove, one bare electric bulb hanging from the ceiling, steel bunks in one corner, outhouses and showers behind. The street already had a name, one that remained for 30 years and was so noted in the phone book—Septic Tank Road. Many of the families that lived in the tiny cabins the first year were able to buy Harris Heights houses at the phenomenal price of $1,000 to $2,500, with Franklin Roosevelt's New Deal mortgage offering through the Federal Housing Administration (FHA) in 1934. But with the business growing, the small cabins continued to be occupied.

Some of Clyde's shacks would house the elderly long-term, those who simply couldn't build up a nest egg for retirement. So in later years Sunshine Bands from the local Adventist church organized groups of school kids and faithfully paid Sabbath afternoon visits to sing for the shut-ins.

My first introduction to this aspect of the mill came in nudges from my schoolmates as 10 of us squeezed into the tiny living room of an elderly woman who must have had the kindest heart in town. Her fastidious husband, long since tired of her dubious housekeeping and stubborn ways, had departed to a trailer beside the tumbledown, added-onto shack. The woman's ancient mother resided on the couch. The aroma proved she had not been moved for a very long time. She was fussed over by the woman of the unsavory manse. But another clue to her benevolence was the kitchen. Beyond the doorway we could see a window over the kitchen sink that opened to a ramp leading to the backyard. This allowed, or rather invited, chickens to cluck their way up the ramp and hop onto the kitchen counter for a feast of cracked corn strewn in their behalf.

In this February foray into the cold to sing for the "less-fortunate shut-ins," I witnessed firsthand an extreme animal lover. It clashed with my upbringing but stirred my conviction that animals had a right to be warm and happy too. And undoubtedly these chickens were the happiest

I had ever witnessed, although my experience with farm fowl was limited.

It was an open secret that my impossible wire-fox terrier, Checkers, found delight in breeching our neighbor's chicken pen, helping herself to unfortunate fowls, one at a time, and proudly depositing them DOA on our highly visible front lawn. She wasn't trundled off to the pound by my parents, because George sent her. (But that's another story.) Together with Checkers' murderous behavior, a vivid early memory of a chicken's vigorous death from having its neck wrung in a friend's backyard solidified my vegetarian leanings. Witnessing the happy fowl on Septic Tank Road that Sabbath afternoon convinced me that kindness to animals came with a certain sacrifice. But after pondering, I decided that Nuteena (until Fri-Chik came along) and a secure, cozy chicken coop would suffice.

WORLD WAR II—PENDLETON AND THE MILL DO THEIR PART

CHAPTER 12

The move to Pendleton came as America was sliding toward another war, but in spite of the world events Clyde and Dette Harris plunged ahead. In 1933 radio stations broadcast Franklin Roosevelt's inaugural address—"We have nothing to fear but fear itself"—as the country withered in the last gasps of the Great Depression. Soon after the May 28, 1940, Harris Pine Mills' grand opening celebration, the deep resonant voice of Winston Churchill boomed into living rooms, "We shall fight on the beaches, we shall fight on the landing grounds, we shall fight in the fields and in the street, we shall fight in the hills, we shall never surrender." Soon America joined the world at war. Clyde Harris was named to the War Resources Board, as the mill readily complied with the government's request for ammunition boxes, and the mill retooled for production. Lumber was shipped worldwide to build military bases. All across America businesses directed their resources into war-oriented items. Many families willingly sent their young men into harm's way and prayed the war would end quickly.

After the Japanese attack on Pearl Harbor, the U.S. Department of War sought retaliatory measures. Lieutenant Colonel Jimmy Doolittle, a retired stunt pilot who happened to be an aeronautical genius, with a doctoral degree from the Massachusetts Institute of Technology, was chosen to strike back at Japan. He discovered that the Army base at the Pendleton airport was training B-25 pilots on the short runway—just what he needed in his experiment to take off and land on ship decks. The men trained at Pendleton to fly B-25 bombers became his special unit of secret forces, and in June 1941 the 1,343-member 17th Bombardment Group was assigned to the Pendleton airport. Barracks dotted the bare landscape on the hill above the town.

Although the war was being fought a world away, it was very real to Pendleton's citizens. Its skies hosted what would become Doolittle's Raiders diving and taking off from the short but substantial runway at the

Pendleton airfield. The townspeople became accustomed to mock air raids held at night and secret missions whose purpose they could only guess. Blackout periods followed. America was on alert in case of an enemy strike. Every town had its rationale as to why they might be a target of the enemy. Heavy window shades were required after dark, and only a minimum of household light was allowed. Each neighborhood had a night warden to oversee light seepage. He monitored his assigned blocks on foot, and if the tiniest slit of light shone through a window, he knocked on the door and politely asked that it be covered. Authorized drivers, their automobiles' headlights taped to allow only a slit of light, strained to see the road.

The Pendleton forces surveyed the coast for Japanese submarines and thwarted a few enemy runs up the Columbia River. Pendleton residents, in fact, were proud of their war status. The Chamber of Commerce posted a sign on Pendleton's outskirts, "Doolittle Raiders Trained Here." The Army soon insisted the sign be removed.

In 1944 the Army formed the Triple Nickels, a unit of Black paratroopers designated to the 555th Airborne Battalion. Lesser known, because they functioned in secrecy, the troops kept the skies above Pendleton lively long after Doolittle and his Raiders completed their 1942 secret mission, taking off from the U.S.S. *Hornet* for a surprise attack on Tokyo in retaliation for Pearl Harbor. The Triple Nickels practiced their jumps in the skies over Pendleton. Sunny Sabbath afternoon potlucks at the Applegate home, with its sloping back lawn overlooking the opposite hill and the airport, proved perfect for sky watching. White parachutes billowed above as the Sabbath crowd cheered the jumpers' perfect landings.

At school imaginative little boys talked in excited whispers about the paratroopers and yearned to fly like the soldiers garrisoned in the barracks at the airport. When a P-38 pilot made a snap decision to show off to his flying buddies, leaving formation to fly low and under the entryway roof of the civilian airport, all of Pendleton heard the news. "Daredevil!" the newspaper fumed. Secretly, however, the boys had a new hero.

The Japanese discovered that balloon-lofted incendiary devices that caught a certain eastward jet stream could reach American shores in about three days. From May to October 1945 more than 200 Triple Nickel paratroopers served as members of project Firefly, an effort to fight the forest and grass fires created by bombs carried across the Pacific by Japanese balloons. These curious devices landing in heavily timbered areas caused great damage.

BESSIE THE STITCHER

CHAPTER 13

With World War II came Norman Rockwell's poster girl Rosie the Riveter. But the meek telephone operator who posed for Rockwell's painting of real life might not have recognized her altered self as a bulked-up coveralled steel mill worker flexing her bulging muscles, bologna sandwich in her fist, and a greasy riveter in her lap. The May 29, 1943, cover of the *Saturday Evening Post* became the symbol of the new woman in the workplace.

The war marked a seismic change for women. Across the nation Rosie represented women in a man's world of factories and became the feminine icon of change. It was the war effort that gave the official nod to the sacrifices of working women everywhere, though at Harris Pine women had worked alongside men for years. Harris Pine women always were equal breadwinners alongside their male counterparts. They evoked the Frank and Ernest cartoon caption about dancers Ginger Rogers and Fred Astaire: "Sure, he was great, but don't forget that Ginger Rogers did everything he did, backwards and in high heels."

Mothers kept the Harris Pine Mill families knit together. Long after their husbands and children were tucked in and fast asleep, these mothers finished their household duties, trudging to the basement to do laundry and ironing for the next day and, no doubt, sometime shedding a lonely tear or two.

That Harris Pine thrived year after year was in no small part a result of the dedication of the female workers who banded together in a feminine fraternity. They endured sickness, loss, death, infidelity, and always money that didn't stretch quite far enough for comfort. Seventh-day Adventist school tuition wasn't cheap, and most families had a child or two and counting. Paychecks turned anemic in the face of rents or mortgages, utilities—no small bill in winter—and keeping the family clothed. In an era before the 401(k), some added to their savings accounts and, when the Harris Pine Credit Union was established, regularly paid into it with payroll deductions. But it was the happy exception that HPM families could systematically create a sizable nest egg for retirement.

With men and boys off to war, women were more and more in demand to fill job vacancies. Not all of them bid their sweethearts goodbye, but many did. Whether their men were at home or at war, the women of the mill workforce bonded, traded stories about their kids, and wished for more help at home or more letters from the war front. Few complained. Before the church took the reins, it was a six-day workweek. With the demands of war, often the boxes sliding down the ball-bearing belt were meant to contain not apples but ammunition destined for overseas. Women buckled down and worked more hours in the day—and more than one mother endured on four hours of sleep a night and liberal use of Empirin 4.

Shipyard poster girl Rosie the Riveter spinoffs such as Wendy the Welder, Julie the Janitor, and Josephine the Plumber could just as well have been Bessie the Stitcher or Lilly or Marie or Clara. Maybe Bessie wasn't holding a bologna sandwich in her fist, but her muscles grew in proportion to the gathering, nailing, stitching, stapling, and stacking of box shook and ammunition boxes.

Bessie Freeman McGuire grew up in the Milton Valley. Her mother and Clyde attended the same school and grade. Somewhere buried in an old family album is a photo of school kids lined up for a picture. It's easy to spot bright-eyed little Clyde in the middle of the front row, but it's not so clear which little girl is Bessie's mother, little Nanette, Nettie Brown. As did so many others in the valley, her family had migrated across the plains. Bessie's grandfather was the first White boy born in the valley, and he gained distinction by putting together the first electrical generating system and selling it to the City of Milton for $500, then operating it for $50 a month.*

Bessie married Bud Freeman and the couple began working for Milton Box in the mid-1930s. Her man didn't go to war, but at home he fought a battle just as deadly. Bessie soon had four children and a husband who was losing his war with booze. She was a genius at multitasking and might well

* Annie Charnley Eveland, "Fabric of Centenarian's Life Interwoven in Legacy of Quilts," *Walla Walla Valley Weekly*, June 5, 2012, p. C5.

claim the prize for holding her family together in the face of greatest odds. The problems she endured were a composite of the many women who slid into their jeans every morning, slapped mayo and Velveeta cheese onto slices of bread, made sure the kids ate their Ruskets, and made a beeline for the HPM front gate before the 7:00 a.m. whistle blew.

Bessie's mother-in-law lived rather than preached the Adventist faith, and Bessie gained comfort in Jesus and joined the Adventist Church. When her husband was sober, there wasn't a better father and husband in the world. His promises, abounding with regretful apologies, were so convincing they seemed etched in gold, and Bessie believed in him. At times her mother-in-law would wrap her in her arms, letting her sob while Bud raged. She prayed with Bessie, encouraged her, and, best of all, provided a wonderful role model for her children. "Please don't leave Bud," she'd plead.

Bessie wasn't the only one who saw Bud's goodness. One evening after supper Clyde Harris pulled up in front of their Milton house and indicated he wanted to talk to Bud out in the car.

"Bud, you're one of my best workers at Milton Box," Clyde told him. "We'd like for you to move with us to Pendleton. I'm offering you the job of foreman of the assembly plant. But there's one stipulation. You have to give up booze."

Bud was sure this incentive was just what he needed to stay off the bottle. He promised Clyde he'd stay sober. "In the meantime," Clyde said, "you'll be in charge of the night shift but without the title of foreman. When you're off the hard stuff, you'll get the day shift and the title."

Bessie and the kids were thrilled, and they moved to Pendleton with high hopes. It was a new beginning, but the booze kept winning. Because of the sporadic drinking sprees, Bud never did achieve the title of foreman, and Bessie took in boarders and laundry for enough money to feed her kids. With her earnings she bought a washing machine to enable her to work even more. She grieved over her husband's mercurial moods, tried to protect her kids, and kept them in church school with the tuition always paid up.

When Bud introduced a surefire plan for his sobriety—the Yakima Valley Klickitat dairy farm, owned by his sister and her husband—Bessie had another reason to hope. The outdoor work would do Bud good, would help him quit drinking. If she would just sell everything, including the new washing machine, and give the money to him, they'd have just enough to move.

"I'll stop drinking. I promise," he pleaded.

How could she refuse?

So with her husband's pledge, Bessie sold her possessions, including the washing machine, and gave all $200 to her husband. They piled into their old car and were on their way to Klickitat and a new start. But first Bud made one stop—the liquor store. It was just for the road, he vowed. Then they were off.

"Bud never made good on his promise," Bessie says, "but some of the happiest years of my life were paling with Bud's sister Nita." They became best friends. "In ways beyond understanding, the Lord gave us the gift of a good life on that farm, even though I was struggling. It was there that I decided not to depend on my husband but on the Lord for happiness."

Sadly, they had to move back to Pendleton so Bessie could go to work again. Bud was very ill and losing his long struggle with drink. He had always been a chain smoker, and now he had lung cancer. At the age of 44 he died.

Under great odds, with fortitude and faith, Bessie endured. She continued to build her sterling reputation at HPM. She was steady, smart, and a hard worker. In 1965, when HPM built a retail store in Pilot Rock, none other than Bessie the Stitcher was tapped from assembly to manage the operation. After all, she'd had plenty of training in the home trenches; as CEO of her home operation with all its problems, she was making but a short step into Harris Pine management. For 20 more years Bessie supervised Pilot Rock Harris Building Supply. "I worked until just four months and 10 days shy of my eightieth birthday, including almost five years after the mill was shuttered and the store changed hands," she says brightly with not a shred of lament. Harris of Pendleton was the grateful recipient of Bessie's intrepid dedication and skills.

YEARS OF PLENTY

CHAPTER 14

After Clyde and Mary's twenty-fifth wedding anniversary and the tremendous push to move the company and build, they could afford to invest in a little pleasure. Together they'd weathered tent dwelling and crises galore, surviving on what they could grow and preserve. That is, except for the 50-pound sack of flour sitting in the corner of the tent, which cost about 68 cents, but then prices had gone up. Even though the lean years seemed far behind them, neither dreamed they would one day own a wheat ranch of their own.

Mary wanted horses. Tennessee Walkers, horses bred for the gentle Southern way of life, were nonexistent out West. She was proud of her Southern heritage passed down from her mother, Susannah, and as Mary studied the smooth-gaited breed, she dreamed of owning one. And so it was that Clyde proposed they take a road trip to the South to look at this unusual horseflesh.

With niece Margaret along, Clyde and Mary drove to Arkansas, passing through other Southern states on the way. It was a lovely trip, made more comfortable in Clyde's 1948 Cadillac. Margaret would recall how thrilled Mary was when she first laid eyes on the coveted breed. Sleek chestnut or black coat, their medium size set them apart, and their gait was smooth as silk. Mary climbed astride, her eyes shining. How could she choose her favorite?

The renowned breeder took Clyde to the stable and introduced him to a pint-sized black stallion sired by legendary champion Midnight Sun. He suggested that Clyde might consider this an investment, being the first to introduce Tennessee Walkers to the Northwest. The colt was nimble, gorgeous, and practically speaking, Clyde lost his head and his heart. Mary's dilemma as to which one to choose was solved. The young stallion and a boxcar full of pregnant mares were soon on their way to Pendleton. To Clyde's way of thinking, purchasing one or two horses was a hobby. But a carload—now, that was a business!

The *East Oregonian* got wind of the purchase:

"Clyde Harris of Pendleton has purchased 15 head of registered Tennessee Walkers from recognized breeders in Kansas, Arkansas, Missouri, and Tennessee. The lot will be shipped to Pendleton within a month or so. . . .

"The collection . . . includes five mares, seven female colts, two geldings, and one stud. The [stallion colt sired by Midnight Sun] set Harris back a cold 10 grand, and a female colt by Midnight Sun, named Sun's Charm, and now 11 months old, cost Harris $2,000. . . .

"All the mares which Harris is bringing west have been bred to Midnight Sun, who by the way was national grand champion (and therefore world champion of that breed) in 1945 and 1946. Harris contends no other horse of that breed has duplicated Midnight Sun's achievement."[1]

But the newspaper didn't let it go at that. In her Random Reflections column Elsie Dickson "was disappointed because the Clyde Harris hosses [*sic*] are coming from the South by train, when everyone knows they are WALKING horses," and she added a cartoon of Clyde and Mary's latest venture.[2]

Clyde hastily built a stable at the back of the mill. Mary installed her brother Frank as caretaker of the carload of prize horseflesh. As the mares produced foals, Frank was a busy man. Now Mary spent much of each day at the stables grooming the magnificent steeds, and soon she would participate in the various roundup parade events. Photos of Mary sitting astride her chestnut mounts waving at the parade crowds remain as testament to the horsewoman Mary became.

It is interesting to note that the horses found greener pasture in 1952 in the Wallowa Valley when Clyde purchased a 1,200-acre wheat ranch. Another of Mary's brothers, Harold, and his family became caretakers of the ranch and the happy Tennessee Walkers. But as for profitability of his venture into horseflesh, history is certain that Clyde was a genius at the lumber and furniture industry, and the horses gladdened Mary's heart. But they never did pad Clyde's pockets.

---

[1] *East Oregonian*, c. 1948.

[2] *Ibid.*

NOW, THAT'S AN IDEA!

CHAPTER 15

Sometimes it's a small thing that sparks a divine direction. The Blalock summer cottage was that nudge. Mary needed furniture. And Clyde was forever indebted to Mary's intuition and the suggestion, for it revolutionized the mill from lumber and fruit and ammunition boxes to an empire far greater than Clyde had ever imagined.

The cabin Clyde built on the flat meadow at the top of Blalock was cozy if one huddled close to the roaring fire. Clyde's rock fireplace was awesome, specifically built to his specifications—created from submerged rocks he had dragged out of swift-flowing icy mountain streams and personally hauled to the Blalock Mountain site.

But even with a blazing fire the log cabin was still drafty. Yet it was Clyde's refuge when burdens became too great. The mountaintop, the forest, the whispering gentle breezes—Clyde needed nothing more. But Mary did. Maybe nothing fancy, but she needed a few of the comforts of home. She knew that if left to Clyde and his brothers, the cabin would remain mighty bare and uncomfortable. It was a tall order, those comforts, considering that the men, swathed in long johns, woolen pants, mackinaws, and boots, tried to outdo one another by hauling down the first and biggest elk each hunting season. Indoor activities centered at Clyde's fireplace, the men cleaning and oiling their rifles while warming their backsides.

Curtains Mary could make, but the place needed some more furniture, so she hired a carpenter to build a couple of bedsteads. The bill was too expensive for her. She reasoned that with all the wood zipping though Harris Pine machines to be made into boxes, why not make furniture, too? Arguing for the thousands of servicemen coming home and so many families coming out of the Depression, she said to Clyde, "I think unfinished furniture would be very marketable." What an understatement in light of the years to come!

So Clyde put some of the factory workers on the furniture project. First, they made a chest of drawers, then a stool, a chair, and a table. Cash-strapped postwar families needed affordable furniture. The box factory began taking on a different look as the new pieces rolled off the assembly line. And the Blalock cabin was transformed to comfortable as Mary arranged her HPM furniture and placed her handmade rag rug, a distillation of her old winter coats, in front of the fireplace.

Mary had her furniture. The nation would soon follow suit. Furniture sales shot through the roof. Clyde jumped into designing and constructing the furniture, inviting Reggie DeVice, a young enterprising manager of a furniture factory at Thunderbird Academy in Arizona, to help sell it. It was a brilliant move on Clyde's part. DeVice put Harris Pine furniture on the map. The orders flowed in, and unfinished furniture swept the country.

Three years after the first furniture slid down the assembly belt in December 1945, Clyde was heralded as the largest manufacturer of unfinished furniture in the world. "Harris Line of Super Pine" became the recognized slogan for the gleaming ponderosa pine and Douglas fir products showcased in virtually every department store in the United States.

"The first articles of furniture made at the new furniture factory constructed at the Harris Pine Mills have been made, to serve as models for volume production to come, and the product—of ponderosa pine from the Harris holdings—is of a quality that will compete with any of its kind in the entire nation. . . . Ponderosa pine will be used, but later the plant may put out some knotty pine furniture as well.

"Our plan is to duplicate our present line of furniture with 'A Beautiful Line of Knotty Pine'—which is our slogan," the manager (R. J. DeVice) said.[1]

In 1954 during the General Conference session San Francisco *Examiner* reporter Mary Crawford asked the couple how they got into making furniture, and published their story:

"We moved to Pendleton, Oregon, in 1939. That's about 30 miles from Milton, and I bought my brother's interest in 1942. That is, he took over some land and I took over the sawmills.

"Then in 1945, in the fall, we started manufacturing unfinished furniture. That was really her idea," he said, pointing to his wife.

Mrs. Harris explained, "Well, we had a little summer cottage and I

[1] In *East Oregonian*, Dec. 1, 1945.

had a carpenter make some beds for the cottage, and I said, 'Why don't we manufacture some furniture?' So we started making chests of drawers and bookcases, and we were in the furniture business, if you could call it furniture. Actually, they were just big crates. We'd be ashamed now of the first things we made, but they were as good as anyone else made [in the unfinished furniture line]!"[2]

At this point in the interview, I can imagine Clyde the businessman wincing and perhaps nudging his wife under the interview table to let her know she'd said quite enough. Mary was warming to her role as cospokesperson. But the couple charmed the reporters with their earnest, open countenances and story of their faith. As their lives took on a more public role, the couple became more and more adept at facing microphones and answering all manner of questions.

"For several years, making furniture was just a sideline for Harris Pine. But Harris furniture caught on, and the company was soon under pressure to expand. It was either expand rapidly and constantly or get out completely," said Harris, according to Mary Crawford. "We decided we had to meet the public demand for our furniture, and soon we were in an expansion program that at times almost took our breath away."[3]

---

[2] In Mary Crawford, "Convert Pays $12 Million Tithe," San Francisco *Examiner*, June 3, 1954, p. 8.

[3] In Crawford.

THE CAT SKINNER

CHAPTER 16

Before any cement could be poured for the box factory, Ken Jones leveled the ground, working with the precision of a genius. He got his nickname one summer in 1947 while picnicking on Blalock Mountain. He and his family were enjoying Clyde and Mary's hospitality one Sunday afternoon with the Bains and Applegates. It was Clyde's way of saying thanks for the hard work and extensive overtime the three men had endured helping to build the company.

Jones handled a Caterpillar like a surgeon handles a scalpel. He dozed the road to Pearson Creek for Clyde, carving across rocky terrain, teasing the blade up and down, leaving behind a smooth, contoured road. He could shove logs, brush, or forest fire with a careful touch to save the wilderness. There was none better than Jones.

Clyde appreciated the loyalty and hard work ethic of the three men. They represented some of the best of his workforce. Getting an invitation to dine with the Harrises among the fir trees on top of their mountain was a real honor. Bain loved Harris Pine. If Clyde and the company made plans, Bain talked in terms of "we." Applegate, Clyde's mechanic during the war years, made parts when new ones were unavailable during those hard times. And, of course, there was Jones, the irreplaceable cat skinner.

So the families toted the usual watermelons, potato salad, and baked beans to supplement Mary's bountiful spreads. But Ken pondered the food issue and came up with something original and memorable. Lately he had been bragging about his favorite food. Bain and Applegate had scoffed and even Clyde laughed when hearing Jones's boasts. Clyde was thrifty, but Jones's specialty took frugality to a whole new level. Well, Jones would show them.

The group met at the North Fork of the Walla Walla River, the base of Blalock. Clyde picked up his guests in his old four-wheeled drive World

War II lorry. Everyone was in high spirits at the adventure of traveling with Mr. Harris up his own road to the top of the mountain.

The eight-mile road in some sections had a 20 percent grade with switchbacks, so Ken didn't let his wife carry the covered platter. Instead, he toted it on his lap. Group curiosity heightened as Mr. Jones placed his platter on the picnic table last.

Not until the right moment did Jones make a move to his front-and-center platter. Not until after Clyde said the blessing on the abundant spread, followed by hearty amens, did Ken finally lift the lid. He did so with the air of a man who knows himself and his talents. As the Harrises and their guests peered at the bountiful heaping plate, somebody in the group gasped. Maybe it was Lois Applegate or Ellen Bain, both decorous and proper women.

Jones's contribution? Battered, fried, and heaped-up—a platter of chicken necks!

Mirth swept through the group, then loud guffaws. Mary cackled. Clyde hooted, as did Bain and Applegate. Mrs. Jones didn't care. Nor did their kids. Everyone would learn why, if they chose to partake of the chicken necks. Jones had a secret recipe. The most delectable chicken anywhere, so his family thought! Colonel Sanders never did get his recipe—but that's another story.

Whether the chicken necks lived up to their reputation as finger-lickin' was not noted—or perhaps conveniently forgotten by the scrupulous Adventist forbears who live on to tell the tale. The group had a wonderful time that day, made all the merrier by Jones's offering. But most memorable was his new name—Chicken Neck Jones.

Jones never did escape the Chicken Neck moniker. He kept clearing roads for Harris, went to church every Sabbath, and whether he became a vegetarian or not, nobody cared. He'd always be Chicken Neck to his friends. One day Elder Robert Pierson, then president of the Seventh-day Adventist world church, would write about him—but not about the chicken necks. It was about how Jones lived to tell another story that has gone down in Harris Pine Mill legends.

BOARDS ON FIRE

CHAPTER 17

The new furniture factory was having growing pains. Progress brings headaches, and as the mill expanded in the 1940s, union rumblings surfaced. Triggered by a couple of union sympathizers, a disturbing undertow was afoot. First, a notice went up by a drinking fountain announcing an evening meeting ostensibly to discuss working conditions at HPM. Loyal HPM workers, irked by the intent of the poster, yanked it down. But it went up again, then again. Curiosity spread as to who was so unhappy that they would attempt meetings to organize against Harris. At first it was just casual chats with employees, encouraging them to come to a union meeting to discuss organizing.

It was no secret that Clyde distrusted labor unions. Most Adventists were leery of organizers and the intent to abridge their rights with the heavy hand that labor unions historically had imposed and demanded. Clyde viewed the union fomenters as adversaries seeding the factory with unrest, arm-twisting, and meetings in his factory. He simply ignored it.

No headway took place. Few workers had time or inclination to attend the meetings fomenting against Clyde and Mary Harris. Most of the mill workers were grateful to work at Harris Pine. Stubborn and independent, Clyde remained intractable. He believed that aligning with an outside influence would abridge religious freedom, and although he took on a certain risk by ignoring the troublemakers, he never flinched.

Clyde was 16 years old when the weekly Milton *Eagle*, January 4, 1906, reported:

"Caldwell, Idaho, December 30, 1906. Former Governor Frank Steunenberg, of Idaho, was assassinated at 7:30 this evening just as he was passing through the gate leading into the yard of his home in this city. The governor had been downtown and was returning to his home when he met his death. He was alone. Just as he entered the gate, a terrific explosion occurred, blowing his body literally to pieces. . . .

"Owing to the supposition that the assassination is a result of the Coeur d'Alene troubles, much excitement prevails in many parts of the state, as Steunenberg was not the only one connected with the arrest and sending to prison of the ringleaders that came in for a share of the threats that were uttered against the officials who took an active part in quelling the strike disturbances of 1899 and it is feared that others may meet the same fate that befell former Governor Steunenberg."

There was hardly a Northwest citizen who had not heard of the silver mine uproar in northern Idaho, the raw power of the union bosses, the ill treatment of the miners. Newspaper accounts were brief, but the unfair conditions and the brutal methods used by the unionizers stirred farmers and field hands alike. Governor Steunenberg had toughened his stance against the labor unions by announcing, "We have taken the monster by the throat, and we don't intend to let go." President Woodrow Wilson brought in federal troops to enforce law and order at the mines. The labor bosses, although well connected politically, had lost, and Steunenberg would pay with his life. The uprisings and explosions in the North Idaho silver mines created a nationwide furor, and the Steunenberg trial took on historic proportions with the defendants' lawyer Clarence Darrow's eloquence for the accused labor bosses. A man who called himself Harry Orchard had confessed to being the labor union's hit man in the retributive murder of Idaho's former governor Steunenberg.

The hired assassin, Harry Orchard, received the death sentence, and the labor union bosses, thanks to Clarence Darrow, were pardoned. Then, through the ministrations of Steunenberg's widow, and the religious tracts delivered to Orchard's jail cell, Harry Orchard experienced a remarkable conversion. *Harry Orchard: The Man God Made Again,* a book by LeRoy Froom, recounts the story.[1] In an unusual act, the warden granted Orchard temporary freedom to be baptized in the Boise Seventh-day Adventist Church. Orchard's sentence was commuted to life imprisonment. He lived the rest of his life in a small shack behind the penitentiary, his extensive garden supplying fresh produce for the prison. The story took on even greater import to Clyde after he became an Adventist.

Clyde simply would have no part with union organizers. But his

[1] LeRoy Froom, *Harry Orchard: The Man God Made Again* (Nashville: Southern Pub. Assn., 1952).

disinterest riled the fomenters. One cold winter afternoon Clyde was driving Mary up South Hill's steep incline to their home. As they passed a car driven by the schoolteacher's husband, the man waved them down in an agitated fashion. Then, pointing at Clyde, he spat words Mary would never forget—"We'll shut you down, Clyde Harris."

Clyde knew the heated tantrum was the result of so little support gained in agitating against him. The evening meetings were not well attended. The dissident was getting nowhere. Frustrated, the man tried to pull a strike. Nobody was interested. His discontent turned to hate. He thought he could threaten Clyde into a bargaining position.

Wednesday night, January 8, 1947. Walt Applegate, the shop mechanic, had been summoned by the night-shift foreman to fix a broken-down fork lift, a machine essential to moving pallets of furniture parts to various locations around the factory. Walt had a reputation for fixing things around the mill, even making new parts if none were available.

He tromped through the snow from his car to the factory entrance, glad he had worn his heavy boots. A warm Chinook had blown through after Christmas and muddied up the place, thawing the surface just enough that the next freeze turned the muddy quagmire as hard as upheaved rocks. Just beyond the factory entrance loomed the sawmill. It was quiet at the moment, on a shortened schedule, as logging activity had slowed for the winter. Live steam, a necessity in the coldest months, poured into the log pond to prevent a frozen mess.

Walt thought he saw a movement off to his left, behind the sawmill. Was it a man's form and a flicker of fire? In a reflexive move he leaped to the factory's inside corner, smashed the glass case, and flipped on the emergency alarm. From the doorway he witnessed the sawmill explode into flames a mere 50 yards away. Walt staggered back. The blaze licked up odd piles of wood and sawdust under the stilted building, raced up the 20-foot pylons and into the machinery, blowing out the windows that secured the saw filing room, and shot flames into the cold January night.

Factory employees dashed from their machines to help. Firefighters arrived in record time, arcing high-pressure streams from heavy canvass hoses. Water rained onto the roof. But the fire gained strength even as the building was soaked.

From North Hill's lovers' lane, Reba and Lilly, parked in separate cars with their boyfriends, were looking down at the twinkling city lights. "When we saw smoke pouring up from the mill site, we both had the same thought—that we'd left the heater on and caused the blaze," Lilly says. The two girls had worked the day shift doing clerical work in a tiny, uninsulated office above the furniture factory. Lilly recalls that when they heard fire engines screaming toward the mill, "It was an awful feeling that we might be the cause for this alarm. We both raced to the mill and almost ran into each other, dashing to our tiny upstairs office. But it wasn't us."

From her living room window Mary Harris saw the flames ripping high at the mill, and she too rushed down. The spray instantly froze on every surface except the blaze and converted the place into a surreal ice-covered whiteout. Lilly remembers slipping and sliding on slick, uneven ground as "firemen urged Mary and the rest of us to go home and let them work."

Rumors flew that an arsonist had escaped as the sawmill burned, parting the brush along the Umatilla River, angling down to the water's edge, and finally to where no one would suspect a darkly clad man on a deadly mission. "We'll shut you down, Clyde Harris" was no longer just an idle threat.

Clyde hunkered down, determined the union organizers would not win. The next morning he faced the loss and set plans to clear out and rebuild. The box-and-furniture factory was reduced to a 50 percent workforce. Clyde asked his employees to hang tough. The rebuilding would begin immediately.

Mary would recall, "Two weeks later the dry kiln fire started from inside. Burned to the ground."

It was, in fact, only 10 days later, on January 18. When it happened, the gym was alive with people connected with the mill attending the usual Saturday night program, a movie. The 14-bay dry kilns' loads of green lumber were drying silently in the mix of heat and moisture. Instantly, it seemed, a holocaust enveloped the building. Again firefighters answered the alarm.

In the darkened gymnasium the movie stopped and the lights went on. From the platform beside the screen Mr. Wentland spoke into the microphone. "We're asking that all able men go immediately to the mill. The dry kiln is on fire."

Amid commotion the men hurried out, leaving the women and kids all talking at once. It was easy to envision the blaze engulfing the mill again.

Firemen, bracing against their bulging high-pressure hoses, could not stem the destruction. Still Clyde would not be bowed. The unions would not prevail. He would continue creating an industry that provided not only good jobs for working people but also products they could be proud of.

The *East Oregonian* reported:

"Expected date for the completion of the new dry kiln at the Harris Pine Mills will be March 10, Clyde Harris said today.

"Equipment for the sawmill, destroyed by fire January 8, is arriving daily and will be in operation by March 10. The dry kiln, of wood construction, destroyed by fire January 18, is being replaced by fireproof concrete.

"Despite the fire, the box-and-furniture factory have been in operation, with a 50 percent crew at work."[2]

But Clyde did one more thing. Just in case. The morning after the second blaze, Clyde showed up at Cy Galogalie's outdoor and gun store, C Cy B4 U Buy. Cy liked the sale he rang up—three Colt .44s. With the newspaper blaring news of the second blaze at Harris, Cy just gave Clyde a thumbs-up and didn't ask any questions.

Clyde issued each of his night watchmen a new .44-caliber pistol. "If you see anything suspicious, shoot in the air," he said. "I don't want you aiming or firing at anyone."

The revolvers didn't stay a secret for long. In their yards and at school, the kids compared notes on the grown-up gossip they heard at home. The unionizers went very silent. A few quit immediately. When school was out, the teacher and her husband quickly left Pendleton. The topic of the fires took up much of the cold, gray winter. By the time daffodils poked their cheerful heads above ground, the sawmill was ready for action and the dry kiln for drying lumber again.

The tale remained unsolved. But when the brethren came to town, they reaffirmed the Adventist no-union policy. As a gesture of goodwill my father, Charlie, suggested that the mill hire the handicapped son of the downtown labor boss. Always among the company names on the fair practices list was Harris Pine Mills.

---

[2] *East Oregonian*, n.d.

A BOLD MOVE

CHAPTER 18

Barely skipping a beat, the mill rebuilt bigger and better from the cinders up. The *East Oregonian* reported: "Harris Pine Mills Plan to Expand—Furniture Factory has Grown in Little More than Year to World's Largest Which Converts Logs to Finished Product."

From the grand opening in Pendleton on May 28, 1940, the lumber-and-box company was Pendleton's pride. *East Oregonian* reporter Nolan Skiff's article noted Harris Pine's escalation:

"On January 25, 1946, the first carload of unfinished furniture was shipped from the Harris Pine Mills in Pendleton, launching this company's pine products in the furniture markets of the nation. . . . Sometime within the next few days, the one-thousandth carload of furniture will roll out of Pendleton from this same factory—now grown to the world's largest firm that converts logs into finished furniture.

"And demand is so great for more products of the Pendleton company, with its furniture now being retailed in every county of the United States, that Clyde Harris, at the request of Bandy and Yeoman, Inc., national distributors, seriously is contemplating enlarging it to a point where present capacity is doubled.

"Such, in brief, is the short but progressive history of one of Pendleton's newest and most successful industries, an industry that gives employment to 400 men from forest to railroad loading dock, and an industry that has an annual payroll of more than $1,000,000 and which expects its annual output to gross more than $10,000,000."[1]

It was an adrenaline-pumping era at Harris Pine. Clyde faced every adversity head-on and won. Yet as his industry increased, he was feeling his mortality. The business, the needs, and the future perplexed him greatly. And undoubtedly, the Holy Spirit was rapping on his heart. Indeed, he was experiencing heart trouble.

[1] Nolan Skiff, in *East Oregonian*, c. 1947.

As Harris Pine furniture landed in virtually every department store in America, into the catalogs of Montgomery Ward and Sears, the demand outstripped the resources. The catalog giants clamored for more and more pages of inventory, pressing HPM to its extremity. Bandy and Yeoman, his marketers and salesmen based in Chicago, the furniture capital of the nation, pressed for expansion. *East Oregonian*'s Nolan Skiff reported about the president of Bandy and Yeoman, Paul C. Bandy: "'The Harris Pine Mills should have a capacity for 10 to 12 carloads daily,' he [the president] said. 'The company is shipping three to four daily and we are selling six. What we need now most is additional manufacture.' . . . Mr. Bandy said that the business here has grown to a point it could multiply annually. 'This plant is the largest of its kind in the world, where the entire process from log to railroad shipment is in one operation,' he said. 'It is now the largest factory making unfinished furniture in the nation—actually making more than all the others together.'"[2]

Along with expansion and sales Clyde's blood pressure was soaring. "Paul C. Bandy, president of Bandy and Yeoman, Chicago, was a visitor in Pendleton yesterday. . . . [He] discusses in glowing terms his company's experience in distributing the Harris furniture. Incidentally he handles all of it—if McKee's or Stroble's want some, they buy it from Bandy and Yeoman of Chicago."[3]

Bandy was getting under Clyde's skin. Clyde chaffed under the heavy control the salesman wielded to his advantages with Harris Pine. Clyde knew the rumors that Bandy was a heavy drinker. Then there was Bandy's ever-rising profit margin as the Chicago marketer saw the pine furniture become such a marketplace hit. Admittedly, Clyde's health problems were getting in the way of his usual measured decisions. Success had plunged him into a cauldron of his own making. Bandy expected more, yet his marketing price tag was escalating as he garnered in sales of Harris Line of Super Pine. On the other hand, Yeoman was easy to work with, liked Clyde's work ethic, valued his religious principles, and had a son ready to enter the business. How to legally disengage from the contractual agreement with Bandy?

On a Tuesday morning in 1949 Clyde suddenly announced he was quitting the furniture business. Beginning immediately, all furniture factory employees would be laid off. Bandy and Yeoman were notified their services would be no longer needed, to what kind of furor, no one remembers. But the contract Clyde had signed was now null and void.

---

[2] *Ibid.*

[3] *Ibid.*

The following Monday Clyde announced he was back in the furniture business and rehired all his factory workers. He was now shed of the contract with his two mid-America marketers. He hired Yeoman and Son as the new sales team. The terms of the new contract favored Clyde. Factory employees had gotten an impromptu unpaid vacation, to the delight of a few, though it pinched others at the end of the month. But, glad to be back on the payroll, few complained.

One spring morning a Sears executive arrived unannounced expecting Clyde's immediate attention. But Clyde's six-seater Grumman plane was gassed up and waiting at the airport along with his fishing buddies for a trout and salmon trip to Kamloops, British Columbia. Clyde explained, but the Sears official didn't take the hint. Today was the day Sears would renegotiate their contract with Harris Pine.

"Sorry," Clyde said, "I'm leaving town."

"We'll just have to take our business to a competitor," the mogul warned.

"Suit yourself," Clyde said, as the executive trailed him through the gravel to his pickup.

Sears made good on the threat and contracted with a small lumber company about 10 miles down the road from Harris Pine. Kerns Lumber at Pilot Rock was delighted to fill some of the Harris Pine Mills orders for Sears. Their men trotted over to look at the HPM inventory (surreptitiously, of course) and copied various products that would show up in future Sears catalogs. Clyde didn't seem to notice or care. Sears kept Harris Pine furniture in its catalog, alongside the lookalike products. And HPM continued to flourish.

Clyde realized that some of his recent decisions had been reckless and were indeed warning signs that his mental and physical well-being were in jeopardy. His health was a high price to pay for the breakneck speed of success. His brothers, all plagued with heart trouble, had died young. He knew the signs. The last was Dette, his partner, the wise brother who had been with him from the dream's inception. Dette had put the dream to work, at each step figuring the materials, the cost, and the number of employees. It was a huge loss when Dette announced he simply couldn't go on and moved back to Milton. His death shortly thereafter left the burden of what they had created solely to Clyde.

For all of his faults and sometimes narrow dealings, Clyde lived an upright life. He loved the Lord, attended church faithfully, and gave generously. Daily he read the Bible and "the pen of inspiration," Ellen White's writings, the ever-present red leather books stacked neatly by his white leather chair.

Later, when news reporters thrust a microphone in his face, he spoke with certainty that when he came to a fork in the road he always asked the Lord what he should do. Mary stood beside Clyde, a softened image of a woman who had struggled with herself and her fears, and yielded.

But the leap from a firm that was owned and operated independently to a giveaway to a world church is a leap that bears explaining. Clyde kept his qualms close the rest of his life. He didn't convey to others the fears that beset him in the deep of night. He didn't talk about those "if only" fears that tested him, taunted him. If he should suddenly die, what could he show? A multimillion-dollar operation and no executive trained to replace him? Thus, in his reasoning and worry, he crept closer to the yawning chasm between fear and peace of mind. To fling himself over would leave Mary standing on the abyss. Yet, the peace he sought and longed for seemed to come only as he returned again and again to the idea of giving the mill to his church.

It was during one of those restless nights, feeling his body folding under the pressure, that Clyde pleaded with God for help. What kept coming to his mind was *Give it to Me*. But Clyde was a realist. He knew you can't just drop a multifaceted business into the laps of preachers and expect miracles. Talking it over with Mary was risky. Like Abraham about to take Isaac up Mount Moriah, he knew what she would say. The mill was tantamount to her child. She could count on it always being there for her. He felt it was too risky to discuss it with Mary before he had made his decision. *But,* he pondered, *what will happen to the mill if I die?* Mary had little knowledge about the empire. Estate taxes could well demand she sell the company. The dream he had so excitedly shared with her when he unrolled and marked a United States map—his dream for providing assembly plants near Adventist schools where students could both work and further their education—would die with him.

"I believe in God's interest in man and His guidance in human affairs," he said later in an interview. "If we lack wisdom we are told in the Bible where to go to get it. I have faith in such divine leadership and know that God has guided in the operation of this plant."[4]

---

[4] "The Sky Was the Limit," *Listen: A Journal of Better Living*, 2nd quarter, 1953, p. 7.

Indeed, the business had outstripped his "vest pocket." The thought of gifting his business to the church is a testimony to his prayer life. As he despaired over his dilemma, his mind went back to the dark days of the Great Depression, when his hands shook while opening the bills. He always gave credit to the Lord for seeing the company through those hardest years. So again, by early 1951, he faced the Lord in the dark hours of the night and prayed, impressed with a bold move, the only one that gave him reassurance and peace of mind. He looked up and said simply, "OK, Lord. I'll do it. You take the reins and lead. I'm going with this decision." With that he slipped into restful slumber.

"THE PREACHERS HAVE COME"

CHAPTER 19

Clyde never wavered after that night of wrestling when he finally gained peace. He would take the next step. He would go to Portland, sit with the brethren there, unfold his plan, and seek their counsel. He returned to the mill focused on making way for representatives of his church to assess the multifaceted company in Pendleton. His trusted warehouse manager, Walt Bain, would be the contact person between Clyde and the brethren to work out the details.

Clyde snapped on the intercom. "Walt Bain, come to the office."

Walt looked at his watch. It was 7:45 a.m. Walt had been with the company for nine years, starting as grease monkey working in the dirt pit under the log trucks. He had changed tires, squirted grease into the gears, and generally kept the mill rolling stock oiled. He worked his way up to warehouse manager and company purchaser. Now with Mr. Harris's abrupt order he stared at his battered old desk inside the HPM warehouse. It was stacked with invoices and sales slips, as every purchase essentially went through this hub of the mill's operations. Axel grease and sifted sawdust covered every surface of the tiny cubicle he called his office. Every conceivable part the mill might need was in the barnlike building, or required the warehouse manager's expertise on how to order it. Hose clamps, multigrade greases and oils, gasoline and diesel fuel, a well-equipped machine shop of tools. It held Clyde's annual purchase of stockpiled rubber tires, roughly $300,000 worth, for every rolling rig for the mill and the woods.

Walt couldn't think of anything he'd done wrong, but maybe this was as far as he would go with Harris. Clyde did have a low flash point nowadays. He'd been known to fire somebody when his ire was up and rehire the vexed employee the very next day.

Clyde liked and trusted the warehouse manager and the brethren would need his assistance at every turn of an inventory. Walt would be the one he would entrust with this company's new direction.

Grease cap in hand, Walt faced Mr. Harris's desk. He would remember the day as one of those awkward moments with Mr. Harris. You never knew what Clyde's quick mind had hit upon nor whether he would be patient enough to explain himself. Today his conversation was woefully short.

"Walt, I've decided to give the mill to the church."

Now, here was a conversation lead-in Walt hadn't expected. "You've what?"

"I'm giving Harris Pine, all of it, to the church."

"Why?"

"Well, I've been thinking it over. My health isn't so good, and if something happened to me, Mary would be in a fix. I decided a couple of days ago. Went to Portland yesterday to tell the brethren."

"Does Mary know? And maybe you should think it over?" Walt asked, hoping to understand this strange turn of Mr. Harris's mind.

"Haven't told Mary yet. But my decision's made." There was a barely perceptible pause. "Just thought you ought to know."

That was all. Walt shuffled, twisted his grease cap around, and finally said, "Well, Mr. Harris, I'll stick by you."

Walt turned to go as a dozen sensible arguments came to him. But he didn't think he dared argue, and so he returned to the warehouse, his mind reeling.

At day's end Walt bumped through the back door of their rented one-bath, two-bedroom bungalow on 15th Street, dropped onto a kitchen chair, and unlaced his boots. Ellen was at the sink scrubbing potatoes. Two bedrooms were a tight squeeze with two children and both grandmothers from Minnesota now living with them. "Ellen, I think Clyde's been standing out in the sun too long," he said as he stood his greasy boots beside the back door.

"Is he sick?" Ellen asked.

"Well, maybe, but it's not heatstroke. I just think he's taken leave of his senses."

"What happened?"

"Well, he called me into his office first thing this morning. Said he just wanted me to be the first to know. He's giving the mill to the church."

Ellen nearly jabbed the potato peeler into her hand. "What in the world is he thinking? Does Mary know?" Ellen, a devout Christian, as loyal to the Lord as anyone Walt had ever known, wasn't prepared for this.

"He hasn't told Mary yet." Walt, a husband who didn't always give his wife her rightful place as his equal either, shot her a rueful glance. "But said he decided last night. I asked him if he ought to think it over awhile, but he said his mind is made up. Then he was through talking." Walt shook his head. "I was afraid to ask any more questions."

They tried to absorb the news, keeping this bombshell quiet. Maybe Clyde would forget his whim. They worried not only about their own family but also about all the people who relied on Harris Pine Mills for job security. They had come to stay. Most Harris Pine workers did. To Walt and Ellen, this talk of Clyde's was foolishness.

In June 1942 Walt and Ellen rolled into Pendleton from Hutchinson, Minnesota, with the whole family squeezed into their dark-blue 1938 Chevy. There were Donna, 5, Darrell, 2, Grandma Michelson, and Auntie Jen, a war bride. Grandma Bain soon followed. Clyde was hiring and that was good enough for Walt. Harris Pine Mills was an answer to prayer.

For years Walt had put in six days a week as a lineman for the Midwest Utility Company. He climbed poles, answered emergencies (usually in the middle of the night), repaired breaks, and teased out hot wires dangling in midair, often in frigid conditions. He rarely had time to consider his future. Living unattached, he enjoyed the freewheeling single life after hours.

His devout mother was anxious to see him settle down. When she suggested he ask pretty church school teacher Ellen Michelson for a date, Walt shrugged and said, "Sure." Asking Ellen hadn't entered his mind, but he humored his mother. He knew of Ellen, but their social circles didn't exactly collide.

What Walt didn't know was that their mothers were church friends and conspiring to match the unlikely pair. Ellen's mother was equally worried that her single-minded daughter might never marry. Twenty-five-year-old Ellen, her widowed mother's sole support, had little interest in dating.

Ellen and Walter couldn't have been more different. Ellen was tall and decorous, while Walter was debonair and charming. He had dark slicked-back hair, a stocky swagger, and a habit of telling tall tales. He didn't go to church—except once in a while just to please his mother. He thought religion was a bore.

But on their first date Walt was charmed. And the school marm found Walt's breezy manner a tonic for her serious nature. Their mothers congratulated each other, but when Ellen sensed she was being swept away, she called a halt. She wouldn't consider marriage to Walter—which is where their romance seemed headed—not until Walter straightened up his life.

Walter wasn't sure about that. Yes, he had fallen for Ellen. Her quiet grace brought out his finer qualities. But going to church? Praying? He would try to win her without those qualifications. But Ellen was in no hurry.

After a five-year probationary engagement, Ellen knew Walt wasn't faking his changed behavior. He truly was a converted man, and they married in 1935. But the Minnesota utility company didn't buy the story of Walt's new life, how nice it was that he was in church every Sabbath sitting beside Ellen. Walt was needed on the job on Saturdays. The work environment grew more adversarial until finally Walter quit. The Depression was in full swing and their rusted-out Ford showed 250,000 miles on the odometer. Their faith would be put to the test.

In the summer of 1938 their fortunes changed. The couple was enjoying the McCloud County Fair, Walt pushing their tot, Donna, in a kind of stroller. Ellen's back hurt, for she was heavily pregnant with Darrell. But she stopped to write her name on a card for a county fair drawing and absentmindedly tossed it into the grand prize drum.

Exiting the grounds at the end of the day, they suddenly heard the loudspeaker blare, "And the winner is . . . Ellen Bain! Ellen, come to the stand and claim your brand-new 1938 Chevrolet!"

This couldn't be! She was dreaming. But they hurried to the stand, where a smiling official handed Ellen the keys to a shiny dark-blue two-door Chevy coupe. As she shoehorned her ample belly into the driver's seat and the gathered crowd applauded wildly for this obviously worthy couple, she knew this dream was real. Forgetting their rusty old Ford parked in the dirt lot, the family exited the fairgrounds to fanfare and a certainty that things were going to be OK.

Now the blue Chevy could get them to Pendleton. Before this the journey would have been a financial impossibility. Those constant letters from Ellen's brother Mick urging them to join his family at Pendleton! His assurances that Mr. Harris would hire Walt at the mill and Ellen could teach at the church school had been nothing but a pipe dream with a rusty car that might get them to the county line before disintegrating. It took some time to settle their affairs in Minnesota and convince both grandmothers of their decision, but finally they struck off for the West.

Now the insecurity they felt at Clyde's announcement filled the couple with that old familiar fear for the future. But after Walt's strange conversation with him that day in 1951, there was no further evidence that Clyde was acting on his wild idea. Every morning Mr. Harris whipped through the warehouse as if nothing was afoot. Maybe his crazy thinking had vanished.

However, Clyde was moving ahead with his decision. Mary was still in the dark, and he worried about telling her. The mill workers and their families were the children she never had. How could he convey to her the struggle he'd gone through about his future and hers and what to do about it? Once his mind was made up, he couldn't casually announce his decision to her while he buttered his breakfast toast. "Oh, by the way, I've decided that we'll give the mill to the church!"

Mary was a practical, frugal woman. You didn't just hand away your life's toil with an "Oh, by the way . . ."

Indeed, when he finally divulged his plans to Mary, it was as if he'd landed a physical blow. Her security was wrapped in the success and stability of the mill. In all of Clyde's past business deals and the possibilities that something could go wrong, Mary would reflect, "Well, we'll always have the mill to fall back on."

Mary knew that at their deaths Harris Pine Mills was willed to the General Conference of Seventh-day Adventists. But now? This struck at the very foundation of her security. There were tears. And yet, though Clyde stubbornly led his life with singular actions that should have included her, in spite of her fears she ultimately supported his decision.

Then one day the intercom again summoned Walt to the office. Wiping his hands on his pants as he strode through Clyde's doorway, Walt saw a dark-haired young man sitting on one of the Harris Pine chairs along the knotty pine wall.

Clyde pointed to the man. "Walt, this is Charlie Nagele. He's from the North Pacific Union Conference. I want you to show him around and help him with an inventory of the place."

Bain sized up the newcomer—suit, white shirt, a tie, black shoes. *The preachers have come,* he moaned silently. Nagele was in his mid-30s, Walt guessed. He assumed this young man had been assigned the Harris Pine Mills reconnaissance and was expected to give his superiors a report. He sensed the man's reticence to wrestle this industry into the arms of the church without due caution.

When with a wave of his hand Clyde doused the small talk and indicated the interview was over, Walt realized he held the power to make it pretty tough for Nagele to gather data and facts. In a sense, the church was at his mercy. Bain knew every nook and cranny of the mill. With a few well-placed remarks he could foment unrest in the furniture factory. Repetitive jobs livened considerably when a juicy rumor floated by on a sleepy afternoon. But Walt had promised to stick by Clyde's decision. And Clyde trusted him.

As they headed for the warehouse, Walt turned to Charlie. "Well," he said, aiming to give due deference, "where do you want to start?"

"Call me Charlie. And I guess it's your call, Walt," Nagele said with a grin.

Inventory meant counting every bolt and every board. Walt knew he wouldn't be disloyal. This was Clyde's wish, and the preachers were here. Nagele pulled off his tie and stuffed it in his coat pocket. He slung his jacket over a dusty chair, unbuttoned his shirt collar, rolled up his white shirt sleeves, and sat down. In spite of his reservations, Walt thought he might come to like this fellow.

If rumors rumbled around the mill, Walt decided they wouldn't come from him. Clyde could break the news as he wished. Walt's job was to get acquainted with this church representative, and the others who would be sent, and help them produce a top-rate inventory.

They worked that whole day until quitting time. Nagele asked Walt if he'd come back in the evenings.

"Fine with me," Walt said.

Every Sunday morning he was at the mill at 7:00 a.m. The two men spent weeks at the tedious job. During that long, hot summer of 1951 a strong bond formed between them. It would remain in place the rest of their lives.

"IT'S YOUR PROBLEM NOW"

CHAPTER 20

The handoff to the Adventist Church wasn't as seamless as Clyde had first envisioned. He thought that with a flourish of his pen he could hand over his holdings. But auditors, accountants, and lawyers thought differently and advised caution. There were legal matters to deal with. The legal and business minds conferred. Denominational officials hesitated, fearing that if something would happen to Clyde, both the for-profit company and the church might be imperiled. It seemed premature to announce the gift to the public until the legalities were solved.

Setting the facts on paper, my father, recorded the events that took place soon after the September 1951 vote taken in Clyde and Mary's living room:

"The General Conference did not formally announce the acceptance of the company in total from Mr. and Mrs. Harris until some months later. . . . However, it was understood by all parties that it was to become a General Conference entity pending settlement of legal questions that must be resolved by the Harris family and the government and the IRS.

"In this interim period it was deemed by Mr. Harris to be owned by the General Conference. Further during that interim period C. J. Nagele personally held all of the stock of the company, endorsed in blank by the Harrises in a safety deposit box at the U.S. National Bank in Pendleton. The above with the knowledge and consent of W. H. Branson, then General Conference president, and C. A. Scriven, North Pacific Union Conference president. The latter appointed by Elder Branson to be a counselor to Nagele. . . .

"[This] came about because of the ill health of Mr. Harris, who suffered from severe blood pressure problems. His fears were that if he died, inheritance taxes could not be satisfied without breaking up the company. His instructions to Nagele, in case of the death of either of them, was to deliver the stock immediately to Elder Branson, who was to claim it for the General Conference in harmony with their verbal agreement."

From November 1951, when my father took Clyde's place on the worn office chair, until January 1953, when the brethren officially announced the gift, the Harris Pine stock waited in a dark, vaulted safety deposit box in my father's possession until the legal matters were worked out.

Drawing in the denomination was a high-stepping dance that included avoiding mine fields into which all the profits might well have evaporated. As the problems whipped past Charlie like the boxcars full of lumber heading east, he "set [his] face like a flint" (Isa. 50:7).

While Clyde and Mary adjusted to retirement, Charlie wrestled with the problems. Clyde had absented himself, flinging over his shoulder the words "You'll never learn anything if I stick around" as he departed. The broad picture of the gift had been voted upon, but as Charlie replaced Clyde at the mill, the details of ownership and what the Harrises deemed to be still theirs had yet to be worked out.

This man and his wife who gave the mill had suffered the highs and the lows of creating the industry. Soon they were to be hailed as humanitarian heroes. In January 1953 the General Conference announced the gift over the newswires. Their pictures and story were plastered on virtually every newspaper in America and many abroad. Yet Mary still suffered the fear they would one day be penniless. She still took her car to the Harris Pine shop to be serviced and twice annually called the factory bosses to send two or three women to her house to do spring and fall house cleaning. And Clyde, still Little Brother to his family, the one who could inveigle his brothers and sisters—and even his mother—into his investment schemes, never lost that little kid's heart that pumped adrenaline when making another deal. And he would always be a little short on cash.

As the old saw goes, you don't look a gift horse in the mouth. What Clyde decided to withhold from the gift and what he gave was still in process, as Charlie soon learned during those first few months. To question closely the details still in the accountants' hands would seem indelicate at best. The handoff to the denomination had left the accounting office scrambling for firm ground and enough cash to maintain continuity. Clyde had made a sortie through the office, essentially taking all the working capital. To say the office personnel were unhappy and confused would be an understatement. But scooping out the till had been an irregular habit of Clyde's through the years.

The beautiful ranch was irresistible to Clyde and Mary. Set against the Wallowa Mountains, the young green of springtime wheat undulated into wind-ruffled gold as the seasons changed, ripening for harvest. For $200,000 the ranch could belong to the Harrises, although they were in the midst of a massive giveaway. How could they buy 1,200 acres of prime farmland set against a backdrop of the snowcapped mountains, the pride of Wallowa Valley? Gazing across the pastures and wheat fields, Clyde and Mary imagined their horses grazing in year-round mild weather. It was a perfect location to restore his mind and body.

Buying the ranch was possible *if* Clyde withheld company cash from the gift. His reasoning was it would be taking from one pocket and putting it into the other. All of the Harrises' assets were willed to the General Conference for mission work around the world. Using HPM funds to buy a wheat and cattle ranch that would one day belong to the church made sense to him.

"The ranch, located about four miles east of Joseph, is noted as one of the best producing farms in the area. It has produced some excellent grain crops and pasture," the *Wallowa County Chieftain* explained in the announcement of the sale.[1]

With the meandering creek winding through the fields, an octagonal red barn, set against a backdrop of snowcapped peaks of the Eagle Cap Wilderness of the Wallowa Mountains, the ranch was one of the largest irrigated farms in the valley. Postcards of the idyllic scene and octagonal ranch sold in town. It was the pride of the region.

Clyde plowed his energy and cash largesse from Harris Pine into the Joseph, Oregon, ranch. Prairie Creek lazed through the middle of the wheat land, and Clyde set to work almost immediately adding more acres to the picturesque scene. At considerable cost and effort he rerouted the creek around the perimeter of the property, redeeming additional acres for wheat. In this day and age environmentalists might call for his arrest. But in the blush of romance with his new property, so poetic that tourists aimed their cameras at the vistas silhouetted against the clear sky, the town just deemed it progress. It made more pasture for his beef cattle and additional acres for wheat and hay.

It was a heady experience for Clyde, who still harbored the heart of a little boy who dreamed of fishing the finest streams in all of Oregon.

---

[1] "N. E. Beaudoin Ranch Sold to C. H. Harris for $200,000," Wallowa County *Chieftain*, Sept. 6, 1951.

Now one ran right through his own property! "He installed a sprinkling system that would ultimately cover the acreage, four miles of underground mainline pipe and three miles of aluminum lateral sprinkler pipe," the *Chieftain* boasted.

The pretty farm grew lush and vibrant under the new ownership and care. As Mary's sleek horses grazed in the fields, her brother Harold and his family took care of the spread as if it was their own. Clyde and Mary had reason to accept the admiring comments and congratulations from onlookers who may have envied the couple's grand style of living.

On the west side of Wallowa Lake, Mary was designing and overseeing the building of a cottage. The beautiful deep, cold body of water was adjacent to slopes of Clyde's timber. In the 1930s, for a few hundred dollars, Clyde had purchased roughly half of the forested western slope of the three-mile lake. He blazed a road, logged selectively, and was left with lakefront property on which to enjoy the view, fish, and rest. For years Mary had made do with Clyde's idea of an improvised cabin—two side-by-side logger shacks with an eight-foot catwalk between them.

It was a brisk November 1951 Friday. A payday.

Charlie had assumed the company's reins just days before. Nothing appeared out of the ordinary, but Charlie wondered how Clyde had managed such an empire with a system of organization largely filed in his vest pocket, so to speak. The need for more accountability and access to accurate numbers would be Charlie's first order of business.

Then a representative from the accounting department in the form of a dapper, three-piece-suited rotund figure appeared in Charlie's doorway.

"Mr. Nagele," the man said, "we've gone over and over the figures, and there's not enough money in the bank to meet today's payroll."

The stout gentleman was quiet, dignified, and impeccable. Bigwigs across the nation who came to do business with Harris Pine seemed to veer toward his office, assuming he was the top man.

Charlie stared at the man in the doorway. "Hold those checks!" he said. "Don't let them out of the office until we solve the problem."

"I'm sorry, sir. We can't do that. The checks have already been handed out."

Charlie blanched. Insufficient-fund checks would reverberate with a noise that would confirm the naysayers before the church could get a firm hand on operations.

"What do you propose we do?" Charlie asked.

The man shrugged. "I don't know. It's your problem now." He turned and walked back down the hall.

The HPM staff had always been loyal to Clyde. Over the years he built a reputation for being a fair boss. And when he came into the accounting office, it was not seen as irregular for him to take money for his personal use. After all, the company belonged to him. Only now, things were different. There was that vote taken in Clyde and Mary's living room. Clyde had signed over the business—but not quite. The legal limbo was the problem.

Whom to be loyal to now? The denomination? It was an unknown, not within the accountants' frame of reference. The fact that Clyde had come into the office and taken $100,000 reverberated around accounting. But who was brave enough to ask Clyde for details. Granted, a new man now sat in Clyde's chair down the hall, but he didn't know the ropes. Why go to him with the problem? It would be like tattling on their boss. The mill had a crisis on its hands. And no one knew where to turn.

Soon after this experience my father noted Clyde's extracurricular finance in a small personal notepad:

Important Personal: Chas. J. Nagele.

December 31, 1951: Agreement CHH and CJN:[2]

Settlement on items CHH felt was due him to be withheld from gift to General Conference:

- Alexandria Plant $6,000.00
- Notes—HPM $60,000.00
- National Wood Products Corporation $24,000.00
- CHH Personal Account $10,753.00
- Total $100,753.00
- Miscellaneous Chicago Plant $500.00
- Total $101,253.00
- In addition, CHH to get two Cummins diesel engines for power plant use at Wallowa Ranch.[3]

Allow me to leap ahead in this story.

Another November, 37 years later.

---

[2] CHH stands for Clyde H. Harris, CJN for Charles J. Nagele.

[3] This in the author's possession.

It was Thanksgiving Day, and our family was gathered in my parents' living room. We were settled on the durable brown HPM couches made at the Upper Columbia Academy's Harris Pine branch plant, while Charlie, an old man now, began to tell stories. We had no inkling it would be for the last time.

He reminisced that, in light of the facts, it's no wonder that the treasurer's office was left gasping for breath as the church rolled into town. Clyde was reveling in his new farm and the luxury of a new cottage on a lake.

Charlie recalled the pressure on him as he assumed leadership of Harris Pine Mills. The brethren departed for Washington, D.C., and Clyde vanished, savoring his new freedom. With his adrenaline pumping again, he was making deals. My father began his story with the man standing in his office doorway calmly explaining the bank shortage, paychecks handed out, and the fact that it was the church's problem now.

The dapper man, his cheeks flushed because of the crisis, looked at my shirt-sleeved father with unmistakable pity. There was no joint accountability forthcoming as he stood there facing the young church representative who now sat in Clyde's tilt-back chair.

As my dad explained so many years later, it took time to counter the dissonance between the church blowing into town and the employees and the bosses getting used to the fact that the mill they'd given so much of their lives to was now church-owned.

"What do you propose we do?"

The man shrugged. "I don't know. It's your problem now."

It was a pivotal moment in my father's career. Where could he turn? The challenge was thrown down between the old guard and these new kids on the block, the brethren. To say that Charlie was among friends who would sit down and work out a plan to salvage the church's reputation would not be telling the truth.

The truth is that friendships were forged and trust was earned between the two camps, but that happened slowly. Charlie was alone—except for the Unseen Watcher guarding this industry, who now stepped to his side.

Charlie could think of only one thing—get to the bank—and what then? He grabbed his hat and hurried to the center of town, the old stone building on Main, the U.S. National Bank. He had yet to get acquainted with A. L. Powers, the president, although Clyde had introduced them.

Charlie had no option but to explain the insufficient-fund checks that were momentarily bouncing into town. Should he ask for a loan? He had to have a loan. But on what? And immediately? Big amounts need board approval, bankers in concert with the head office in Portland. The alternatives he faced at 11:00 a.m. that Friday were as barren as Harris Pine's bank accounts. Mr. Powers was Charlie's only hope. At that moment the razor's edge of disgrace threatened to topple the gift and the church's reputation. What could be done to save this venture?

A. L. Powers, bank president, listened with a banker's measured calm as the new executive, in deep anxiety, sat across the gleaming mahogany desk with his head in his hands. The problem lying in the gap between them seemed insurmountable.

Mr. Powers leaned back in his chair and reflected a moment before he spoke. "Charlie, I'm going to do something I've never done before. I'm going to trust you and your church. On my own signature I'll give you the loan to cover the payroll.

"But," his voice was grave, "the money better be in the bank by Monday morning, or my job will be on the line."

Forty years later, as he recalled the facts to his family, Charlie's voice broke with emotion. His hair was silver now and he was bent with age. But he recalled with deep gratitude to God for providing a way on the Friday that almost spelled ruin for the gift, the day when the church's reputation was imperiled.

A. L. Powers became his lifelong friend. Testimony to that fact is a treasured letter from A. L. Powers to Charlie and the brethren in Washington, D.C., when years later Powers announced his retirement and reflected on his friendship and trust in Harris Pine and Charlie.

As my father unfolded the story, it was a sacred moment. God was at the center—unseen, unspoken—but we felt the goodness, the solid Christian character of this banker willing to stake his career on a church and a company in transition. It's no wonder that he would soon be called to a higher position and become vice president of the U.S. Bank Northwest.

Lost in the archives of time, the emergency shortfall generated by this newly acquired gift is tucked into the General Conference archives, the record of a business transaction as recorded at the General Conference Committee, November 21, 1951:

"VOTED, That the North Pacific Union be authorized to borrow $100,000 for a period of from three to five months."

For serveral months the mill leaned on the church to cover shortfalls. And then the tide turned. In the first year of church operation the company turned in a gross volume of nearly $4.5 million. Over the approximately 30 years of denominational operation that figure grew to more than $63 million per year.[4]

The money was paid back by Harris Pine Mills, and in addition, by the close of 1952, Charlie handed the General Conference a check for $100,000. God had been there all the time.

---

[4]Charles Nagele, "Bits From the Past: C. H. Harris," Harris Pine Mills, Seventh-day Adventist Church, 1986.

THE VELVET BUTTON—THE MEDIA DESCENDS

CHAPTER 21

By God's grace Harris Pine Mills weathered the crisis. The shortfall was short-lived, and by the end of 1952 Harris Pine handed the General Conference $100,000. But the stock of the company remained in lock box security for almost a year while legal details were being worked out.

On January 28, 1953, church leadership and the Harrises released a joint statement, and the news of the windfall the Adventists received was broadcast across American and foreign newspapers. Clyde and Mary posed for pictures and did their best to answer the questioning media. The mill didn't change during that year, but with the news announcement the public's perception of it did. Clyde and Mary were besieged by reporters asking why. Clyde had a stock answer, and Mary followed suit. "We always paid tithe and the Lord blessed. We did right by God and He did right by us." The following are some of the titles and headlines:

Terre Haute (Indiana) *Star*:
"Thankful Gives His Furniture Plant to Adventist Church"

Minneapolis *Morning Tribune*:
"Man Gives Church Business Firm Worth 10 Million"

Cleveland *Plain Dealer*:
"Gives His $10,000,000 Lumber Firm 'to God' "

Time magazine:
"$5,000,000 Tithe"

On January 29, 1953, the Hartford (Connecticut) *Courant* published an article entitled "Seventh-day Adventists Given Harris Millions":

"Pendleton multimillionaire today gave his multimillion-dollar lumber business to the Seventh-day Adventist Church.

"'We have tried to be faithful in our dealings with the Lord and He has given us far more than we deserve. Now we are giving our entire business to Him,'" said Clyde H. Harris, 62, speaking for himself and his wife. . . .

"The denomination regards Saturday as the Sabbath and performs no work on that day. That fact was a sore point with orchardists in the region when Harris opened his fruit box factory in nearby Milton in 1913.

"'Some of those orchard owners got mighty upset when I told them we just would not work on Saturday. They threatened to stop buying from us, but they never did. Our business went right on booming. You see, we did right by the Lord and He did right by us.'

"More than half of the 600 employees in Pendleton are members of the Seventh-day Adventist Church.*

In the Philippines the headline in the Manila *Bulletin,* "Errand Boy Gives Back a $10 Million Tithe," by B. B. Alsaybar, caused many to stop and read the details.

Following are more newspaper press release statements:

Chicago Daily Tribune:
"Church Given Lumber Plant—
Oregon Couple Makes $10,000,000 Gift"

Pittsburgh *Post-Gazette:*
"Oregon Millionaire Gives Big Business to Church—
Seventh-Day Adventists Receive Lumber Enterprise
Because 'Lord Did Right by Us'"

Minneapolis *Star:*
"Couple gives $10,000,000 Business to Church"

Chicago *Daily News:*
"Millionaire Gives Firm to Church,
Says God Deserves Share of Income"

Chicago *Daily Tribune:*
"Church Given Mills Valued at 10 Million"

---

* Robert D. Byrnes, "Seventh-day Adventists Given Harris Millions," Hartford *Courant*, Jan. 29, 1953, p. 2.

Flint (Michigan) *Journal:*
"Church Given Lumber Plant"

The *Pittsburgh Post-Gazette* announced, "Oregon Millionaire Gives Big Business to Church":

"A Pendleton millionaire today gave his multimillion-dollar lumber business to the Seventh-day Adventist Church.

" 'We have tried to be faithful in our dealings with the Lord and He has given us far more than we deserve. Now we are giving our entire business to Him,' " said Clyde H. Harris, 62, speaking for himself and his wife.

"The business has an estimated value of $10,000,000. It includes the 600-employee Harris Pine Mills here, one of the largest producers of unfinished furniture in the nation. Also included are furniture assembly plants at Chicago Heights, Ill., Alexandria, Va., and Cleburne, Texas.

"President W. H. Branson of the Seventh-day Adventist Church said profits from the business would be used to help support the denomination's churches, hospitals and schools here and abroad."

More headlines:

Dayton (Ohio) *Daily News*:
"Church Given Lumber Firm"

Salt Lake *Tribune*:
"Church Given Lumber Mill"

Detroit *News*:
"Pair Give $10,000,000 Business to Church"

Wheeling (West Virginia) *News-Register*:
"Couple Give $10 Million Business to Adventists,
Oregon Pine Mills Made Gift to God"

Florida Times-Union:
"Plant Worth Ten Million Given Church"

Toledo *Blade*:
"$10-Million Firm Given to Seventh-Day Church:
Oregon Business Man, Wife, Donate Pine Business
Because 'Lord Did Right by Us' "

Macon (Georgia) *News*:
"$10 Million Lumber Firm Given to Church By Couple"

Seattle *Daily Times*:
"Harris Pine Mills Given to SDA Church:
Huge Lumber Business Is Harris Gift"

Danville (Illinois) *Commercial-News*:
"Seventh-Day Adventists
Get 10-Million-Dollar Business"

St. Louis *Dispatch*:
"$10,000,000 Business Given
To Seventh-Day Adventist Church"

Dubuque (Iowa) *Telegraph-Herald*:
"Business Worth 10 Million Goes to Adventists"

Harrisburg (Pennsylvania) *Evening News*:
"God's Share, Man and Wife Say; $10,000,000 Lumber Firm Given Adventist Church"

Chicago *Herald American*:
"Plant Near Chicago—Adventists Get $10 Million Gift"

Duluth (Minnesota) *Herald*:
"Church Given 10 Million Dollar Firm"

The Chicago *Sun Times* reported, "Lumberman Gives Back $10 Million Firm to God: Pendleton, Ore." It stated the following:

"Clyde H. Harris, 62, owner of the Harris Pine Mills, said: 'Let's show the rest of the world how America lives under God. . . .' Clyde and his brother C. B. (now dead) started a paper box business back in 1913 which skyrocketed into a $10 million business. Wednesday Clyde Harris turned it over to the Seventh-day Adventist Church. Included in the gift is a plant at 1644 Center in Chicago Heights, Ill. 'Why wait until we are dead before we give it to the church?' he said. 'After all, we made our money under God, with the aid and guidance of the church we attended since childhood. That's why my wife, Mary Viola, and I decided to give it now and see the money doing good while we are still among the living in this great country.' "

Still other newspapers featured the phenomenal story:

Daily Oklahoman:
"Religious Sect Given Millions by Lumberman"

Fresno *Bee:*
"Seventh-Day Adventists Given Harris Millions"

Joplin (Missouri) *Globe*:
"Business Worth Millions Given to Adventist Church"

Scranton (Pennsylvania) *Times*:
"Couple Gives $10,000,000 Business to Church"

Madison (Wisconsin) *Capital Times*:
"Church Given $10 Million Lumber Firm; 'Lord Good to Us' "

Hollywood *Citizen News*:
" 'Doing Right' $10,000,000 Firm Given to Church;
His Business Is 'In the Lord's Hands' "

Speaking of this unique business transaction the *Oregon Daily Journal* editorial dated January 30, 1953, entitled "A Great Gift in the Christian Spirit," stated:

"Mr. and Mrs. Clyde H. Harris of Pendleton have given a striking demonstration of what happens when people take their religion seriously. They have turned a lumber and furniture business, reportedly worth $10 million, over to the officers of the Seventh-day Adventist Church.

"It is the largest single gift ever made to the SDA organization and certainly one of the largest individual contributions ever made to an American church. We can imagine that the generous Harris family is considerably surprised over the publicity that has followed the gift. It all seems very simple to them! . . .

"Mr. and Mrs. Harris have so many years of life ahead of them that they may continue to count heavily on this pledge. All of us will wish them long life and continued usefulness."

I was 12 years old the day *Life* magazine news reporters came to church. With my winter coat buttoned all the way up against the drafty gymnasium, I attended the service along with close to 1,000 other church members in the Quonset gym Clyde had built in the mid-1940s next to the outgrown sanctuary. This multipurpose facility was used for school activities, church socials, weddings, and church services.

Every Friday afternoon deacons set up row upon row of folding chairs for the Sabbath services. The stage was lined with the same squeaky chairs for the minister and elders. This one Sabbath I sat next to my mother and brother, for my father was on the platform sitting beside Mr. Harris. Elder E. J. Bergman (formerly our pastor who, at the time, had recently become the Sabbath school secretary at the Oregon Conference of Seventh-day Adventists), with the peculiar swing of his wooden leg, went to the pulpit and welcomed the crowd, which included *Life* magazine reporters. The usual church events had an air of nervous decorum as newsmen, cameras swinging from their necks, hauled their tripods to strategic locations. They lurked at the edges of the platform aiming their viewfinders, stalking the perimeter to get a better angle of Mr. Harris, who was slated to give the morning prayer. Various other photographers targeted the congregation. Some nervously puffed on cigarettes. Indeed, to them this was a strange scene. And who was this Reverend Nagele replacing Mr. Harris? *Life* magazine needed an angle—how to deliver this story?

How surprising it is when a small reminder blinks an old memory into focus. Some time ago I remembered the long-forgotten gymnasium scene. I was digging through my tin button box. It had once contained a rare vintage kaleidoscope—from ordinary shirt buttons to the six hand-carved ivory rectangle buttons my father gave to my mother one year when he returned from the annual North Pacific Union six-week Alaskan bush pilot trip. The creamy, smooth treasures and other sublime remnants resided in the box that my grandmother had given to my mother, which she then gave to me.

The ancient lid's hinges were bent. It had been raided of the ivory and mother-of-pearl beauties by my 6-year-old son for an art project. In a harried, unguarded moment I let him take the box to school, and he returned with the modest leftovers after his classmates picked out their favorites.

Years later I held the lone survivor of that first-grade art raid: a softly round ochre and blue velvet coat button—the only remnant of that sturdy winter coat. In an instant I was transported back to the cold February day in 1953. A thousand folding chairs creaked, shoes scuffed the hardwood floors—the sounds magnified in the enormous ceiling. On bad weather days we skated there every recess and noon or dashed to the closet that held the volleyballs.

After church I remember gazing at the hundreds of just-vacated chairs, the worshippers gone to the Sabbath dinners warming in their ovens. My mother had already hurried up the steep steps behind the gym that Clyde had added for his convenience. It was a hardy climb but only a half-block from Clyde and Mary's home. She planned to help Mary with Sabbath dinner, which would be a photo event for the *Life* cameras. Even though the photos taken around that table never made the magazine, the dining room meal is a vivid memory of the Sabbath that *Life* magazine came to dinner.

But even though church had ended, my curiosity held me back. I remember standing by the drafty double doors, the wind chill cutting through my coat, as the excitement wound down. Reporters still clustered around Mr. Harris, my father, and a few others. They were aiming their cameras and scribbling on small notepads. I watched as one reporter stepped away and began nosing into the cluttered janitor's closet just inside the gym's main entrance. It was filled with brooms, volleyballs, nets, and random old skates.

So this was the way news is gathered, I remember thinking. I was fascinated by their lack of propriety. My father was busy with the cluster of news professionals, so he didn't notice. The closet explorer emerged successful, obviously pleased with his pawing through the clutter. Those connected with the Harris Pine story were not equally delighted when they learned how the reporters used the find to spin the story of the gift. Left behind after an evangelistic series was a rolled-up banner that heralded "The Gospel to all the World." With that, *Life* magazine had their angle.

On Sunday the magazine team summoned Clyde to a photo shoot at the mill. Compliant and unaccustomed to media exposure, he willingly posed at the conjured scene. With the bold banner unfurled against the high sawdust conveyor belt, Clyde clamored atop a log pile and, for the magazine's sake, supposedly announced the gift to a gathering of curious mill workers. The truth is the mill folks were already accustomed to the news. My father had been in place for more than a year. The mill was thriving and the workers had settled back, assured that their jobs were not in jeopardy. But the bemused factory workers posed patiently, and some can still proudly point out their image in the grainy magazine photo.

The photographers came to dinner that Sabbath but didn't sit at Mary's heavily laden banquet. They hovered behind our chairs, mostly focusing on Clyde and Mary. I tried to remember my manners, poking delicately at the mashed potatoes and gravy, carefully passing along the heavy silver dishes. I even throttled my notion to ting my sterling fork against the sparkling water-filled goblet for that pure tone of quality crystal.

So many details have dimmed like an old color snapshot found faded into browns and oranges, the color of time blotting out the sharp detail. To my adolescent senses it seemed almost magical as I tried to unobtrusively fit in while absorbing the event. Aunt Mary sputtered to the women in the kitchen about those *Life* magazine fellows smoking in church! She was appalled and said so just out of their earshot.

The story splashed its way around the world. I remember how quietly disturbed my parents were at the mistaken details newspapers printed as truth and at the angle the magazine took. That Sabbath at the back of the gymnasium I saw and heard more than I realized. The sights and sounds of spin and angle had very little to do with the true benevolence of the story and the God who had parted the way for this business again and again.

The world made much of the gift but didn't always respect it. Yet God consecrated it. Not because it was perfect. Not because it came scrubbed clean and wrapped with a bow. It came frayed at the edges, stuff spilling out, the contents not always pretty. It's interesting what God values and where one finds His stamp of approval. It might not be found in the news archives. I placed the velvet button back in the vintage box and pinched the lid closed. The reminder of our mysterious God was not lost on me that day.

CHAPTER 22

SWEATING THE IMAGE

The mill office brought the Harrises' fan mail to their home in apple boxes. The letters ranged from poignant needy stories begging money to silly requests. Mary read every letter, wiping away tears at the distressing appeals for money, laughing at the outlandish. One letter in which she took particular took delight was addressed just to Clyde. The letter writer, a woman, owned a trailer, loved to travel, and, she asserted, was fairly young. She wanted him to run away and marry her. The enclosed photo revealed that "she was good-looking, too," Mary would laugh, adding, "Clyde really enjoyed that letter."

A friend sent them a clipping out of the Atlanta *Journal and Constitution*. In "Moral of This Story Is: Keep Closer Count on All Your 'Kissing Cousins'" clergyman and columnist Pierce Harris briefly, with humor interspersed, described the genealogical background of the Harrises and caught the spirit and intention of the gift:

"All my life, I've said the Harrises are either politicians, preachers, or policemen; and every one of 'em as poor as a church mouse. Now here comes a fellow from Oregon giving his church $10,000,000 and claiming his name is Harris.

"I'm going to take another look at our family tree. I was out in that country last spring and the fact I neglected to visit Cousin Clyde is going to grieve me the rest of my life.

"That's the most expensive 'neglect' I've ever been guilty of. From now on, when I'm in a 'furren' country like the Far West, I'm going to carry my genealogy book along and read it every time I come to a new town. There may be some more Harris cousins sitting on a nest full of golden nuggets and I might pass right by their front door and not know it. . . .

"But now for a serious thought, although that of Cousin Clyde leaving me plum on the outside of his $10,000,000 melon cutting is pretty serious to me. Why do not more people remember their church in their wills?

"One of the most selfish wills I ever read was made by a friend of mine who was close enough to me to ask me some advice about how he would bequeath what he had. When I suggested he put his church in for $50,000, he demurred and said, 'That would not please my family at all.' And it probably wouldn't have—but it would have pleased the Lord. And that is not something to be brushed aside, especially when one is expecting to see Him before long and will be called on to give an account of his stewardship."[1]

As the articles and stories spread, the couple basked in their newfound celebrity status, even as Harris Pine Mills struggled with the image created by the announcement. The media had succeeded in casting a long shadow over the company's new owners, predicting that a church-run business portended a quick downturn and ultimate demise of such an undertaking by "the preachers."

The cryptic news headlines grabbed nationwide attention. Some papers included a picture of Clyde seated at his desk, flanked by Elder Branson with my father. Column inches were spare, but some referred to the church's leadership replacement as the Reverend C. J. Nagele. *Life* magazine, February 16, 1953, reported, "Reverend C. J. Nagele, although he has never had any previous business experience, has been in training for the last year to run the plant."

With the subtle insult, my father faced an uphill battle. The story slant did not favor a successful outcome for the company. Coupled with the headaches of employees' fears and biases, it was on his shoulders to bolster company credibility and establish his competence.

Charlie leaned into the challenge, alternating between despair, petition, and thankfulness. That first year (without a foreknowledge that Clyde would disappear the day Charlie stepped into the knotty pine office) proved more pivotal than anyone had imagined.

In the beginning the till went dry and the company had to be stabilized through its identity change. In addition, the workforce had yet to build trust that nothing would change except that all profits would go to the church. But it was largely Charlie who had to prove it. He was the face of the new owners—the leadership of the Seventh-day Adventist Church.

---

[1] Pierce Harris, "Moral of the Story Is: Keep Closer Count on All Your 'Kissing Cousins,'" Atlanta *Journal and Constitution*, Feb. 8, 1953, p. 4C.

But by the end of 1952 Harris Pine was solvent, Charlie was still standing, and out of profits a $100,000 check had been gifted to the General Conference. The mill was thriving.

Yet a knot of resentment lingered, for when Clyde announced his decision to gift the mill to the church, he circumvented his loyal managers. Now their expectations for upward mobility seemed doomed. With the church now in control, many felt that everything had changed. Happily, they would be proved wrong. With the development of the student work program and the proliferation of branch plants across the United States, employees became strong believers in "the work," as church members called it. Many of the younger generation trained to become branch plant managers and slipped easily into their new church-oriented work.

Outside the company other issues were at stake. Business connections clamored for assurances. Competing furniture manufacturers hovered over the news, hoping a quick death of Harris Pine would be their chance to replace the world leader in the unfinished furniture market.

There was one happy exception during those first months, as told by D. A. Delafield, who reported at the 1954 annual Fall Council:

"Monday evening the Harris Pine Mills showed a beautiful film representing the work of that large furniture-making institution which has been turned over to the church by its owner, C. H. Harris. I was reminded of a conversation I had just a few weeks ago with the general manager of a large furniture store in Washington [D.C.]. I was looking for a piece of furniture for a bedroom in our home. 'Let me get you the Harris Pine Mills catalog,' he said. 'This organization is the biggest manufacturer of unfinished furniture in the world.'

"'Did you say Harris Pine Mills?' I asked.

"'Yes,' he replied.

"'Did you know that Mr. Harris is a Seventh-day Adventist and that he recently turned over his business to the church?' I asked.

"He looked at me in surprise.

"'Do you mean to say that Seventh-day Adventists operate the Harris Pine Mills?'

"'Yes,' I answered, 'with the guidance and counsel of its founder, Mr. Harris, who is a loyal supporter of our worldwide work.' It was thrilling to see the change that came over the gentleman's face as I spoke to him.

"'Jack,' he said, calling his associate, 'did you know that the Seventh-day Adventists operate the Harris Pine Mills?' The expression on this young man's face was one of great surprise. The look on their countenances revealed the unspoken thought—*I never knew that the Adventists had anything like that!*"[2]

But in contrast to the tale of the furniture store manager, when the news reached the desks of big accounts loyalty to Harris Pine Mills wobbled a bit. It was a skeptical climate that faced Charlie and his associates in 1953 as the press notice went out. The big accounts (Sears, Montgomery Ward) and, of course, the salespeople from Chicago and points west and east, north and south—few knew anything about the Adventist Church other than the rumors that it kept to strict rules and wouldn't do any business on Saturday. Charlie's first challenge was to tackle the hurdles, gain employee trust, develop friendships with wholesalers and retailers, and prove to the banks that Harris Pine Mills would remain solvent.

Back in those days Trans World Airlines and United Airlines had cache and people dressed in their best to fly the friendly skies. Charlie's itineraries took him on United's DC-6s six hours overnight to Chicago, going east. The plane hugged the prevailing winds to enable Charlie to arrive on time for morning meetings where he faced skeptical executives.

A preacher? This they had to see.

It took courage Charlie didn't always possess as he pressed the elevator button to the executive floor and heard a secretary murmur into her boss's intercom, "Sir, that preacher from Harris Pine Mills is here to see you."

But humbling as it was, Charlie gave his best speeches, countered when the talk turned to the whys and the wherefores of church ethics, and made the most out of the charming Harris couple who were at the center of the gift. The moguls respected Clyde Harris. There was general admiration for a man who had wrested apple boxes into a world furniture industry. But it was more than just the preconceived notions of the public that concerned Harris Pine. If those biases toward the company hindered sales, it would cancel accounts and slow marketing efforts. Harris Pine would be in for rough water.

As executives across the nation listened to Charlie, within them stirred a conviction they did not fully understand and in ways they had never encountered in business before. Their hearts were pulled with admiration at Clyde's "leap of faith," as they saw it, at the same time impressed with

[2] D. A. Delafield, "Harris Pine Mills," *Review and Herald*, Nov. 11, 1954.

the company's new goals and mission. One of Charlie's first hurdles was Montgomery Ward's legendary leader, Lee Sewell Avery.

GLOOMY SEWELL AND WARDS

CHAPTER 23

It was a fine Chicago Sunday afternoon in 1931 when Lee Sewell Avery, enjoying his golf game, noticed a contingent of men striding toward him. Sent by J. P. Morgan, Jr., the major stockholder of Montgomery Ward, and the company's banker, a few of Ward's board had been commissioned to confront Avery with a business proposition. The mega company was on the ropes and in dire need of Avery's grumpy financial magic. Dubbed Gloomy Sewell by his friends, he was at the top of the business world—president of U.S. Gypsum, a plaster and wallboard company created largely by his father. Even though Sewell ran his business as if the sky might fall tomorrow, he didn't lack wealth or social connections.

Cornered on the golf course and presented with a lucrative offer to head up Montgomery Ward, for once Sewell didn't shoot off his mouth but took the polite way out. "I'll think about it," he said, with no intention of taking the position. Lucrative offer aside, his position at U.S. Gypsum was more esteemed and, to put it mildly, the thought of heading up a mail order business that sold to farmers didn't interest him.

Avery was in no mood to take the helm of a "yokel" company started in 1872 by traveling salesman Aaron Montgomery's one-page handwritten list of his products and direct mail sales to country people. This "catalog," it should be noted, would one day become the universally known Wish Book.

But after the encounter on the golf course the group telegraphed Morgan, or, some say, Avery himself sent the telegraph. It was claimed the transmission was garbled. Nevertheless, Sewell awoke Monday morning to newspaper headlines announcing 58-year-old Sewell Avery as the new president of Montgomery Ward. He was mad! But now that it was a matter of public record, he begrudgingly joined Wards to save face but kept his position as president of U.S. Gypsum. Yet he never missed a chance to leak his disgruntlement. "I banana-peeled into this place," he'd complained, "and then couldn't get out."

Nevertheless, in his first three years with Montgomery Ward, the "company went from a nine-million-dollar loss to a nine-million-dollar profit."[1] He became a business legend in the midst of the Great Depression. But he continued to gripe at being railroaded by his friend J. P. Morgan, Jr., and the contingent of big shots who interrupted his Sunday afternoon golf game taking his polite evasion for a yes.

In a description that does not stretch the imagination to summon the persona of Clyde Harris, a 1936 *Fortune* magazine opined that Avery was "pretty generally held to be the No. 1 Chicago businessman. . . . Picture a man of average height, . . . thick brown hair, iron gray at the temples, and the deepest, brightest eyes you have ever seen—eyes that show you how fast the brain behind them works. . . . He represents that peculiar combination of qualities that marks the successful Chicagoan—he is shrewd, frank, honest, persuasive, hard-boiled when he needs to be, sentimental when he pleases to be."[2]

He launched his new career with the irascibility for which he was already famous, grabbing Wards by the ears, and he didn't blink as executives and employees leaped off the ship. He once told stockholders, "If anybody ventures to differ with me, of course, I throw them out of the window."[3]

Before Harris Pine began making the furniture that so delighted Avery, he had defied a direct order from President Franklin Delano Roosevelt demanding that Wards sign a wartime union agreement to fire any employees who wouldn't join the union, or risk government takeover. Using wartime emergency powers, Roosevelt ordered the Army to take over Wards. Avery remained firm. He would not allow outside forces to dictate whether his company would unionize. He simply ignored the presidential order, sitting at his desk every day in unyielding composure. News photographers captured the unyielding tycoon:

"The actual ejection of the $100,000-a-year board chairman of Montgomery Ward was carried out. . . . The two soldiers picked 170-pound Sewell Avery up by his arms and thighs, carried him to the elevator. Mr. Avery refused to walk in; the soldiers picked him up again. . . . On the main floor he again refused to budge. The soldiers hoisted him up, carried

---

[1] James Grant, *Money of the Mind: Borrowing and Lending in America From the Civil War to Michael Milken* (New York: Farrar, Straus & Giroux, 1992).

[2] *Ibid.*, p. 26.

[3] *Ibid.*

him past a handful of startled clerks in the lobby and down the main steps. His grey hair ruffled, his blue suit coat buttoned, his hands folded benignly across his stomach, his eyes half-closed, Avery looked every inch an Oriental potentate being borne by slaves."[4]

The soldiers set him down. He ambled across the street to his waiting limousine. The government takeover didn't last long. With legal wrangling over the constitutional right of the government takeover, the Army withdrew after four days, and Avery was back in office and his green leather chair. They accomplished nothing in the four-day takeover. Avery went back to work.

Avery ruled, like it or not, for 23 years, selling Harris Pine Mills furniture with a zest. In a stockholders' meeting he once bragged, "We no longer depend on hicks and yokels. We sell more than overalls and manure-proof shoes."[5]

And so Avery began distributing Harris Pine's furniture. Clyde Harris and the Chicago tycoon weren't too far apart on values. Raised a strict Presbyterian, Avery didn't differ very far religiously from the mill's founder. They both loathed labor unions as an abridgment of rights. Avery was in full accord with the Adventists and Harris Pine. He liked the man behind the wildly popular furniture product, and by the time the church arrived, the Harris Line was vying with bedding as the top supplier of products to Montgomery Ward. Given the large shadow Sewell Avery cast over the business world, Charlie knew that if Montgomery Ward deemed Harris of Pendleton a winner in spite of the church takeover, all the other big accounts would follow. Yet, who could be sure how Avery would respond? His spot-on hunches could make or break the company.

Soon after the church assumed the Harris Pine reins, Sewell Avery, who counted his friends among the richest business leaders in America, found himself seated beside Charlie at the company banquet. The special occasion was to honor the top Montgomery Ward's suppliers. Charlie had come to Chicago to accept the honor on behalf of Harris Pine Mills.

Charlie didn't love banquets, and in the first few minutes at the table realized that Mr. Avery didn't either. Despite the soft, smooth sounds of

[4] *Time*, May 8, 1944.

[5] quoted in "Mr. Avery's Ex-Men," *Time*, May 6, 1940, www.time.com/time/magazine/article/0,9171,763972,00.html.

a live symphony orchestra and the glittering surfaces and surrounding mirrors reflecting the bouquets of flower in the candlelight, Avery, now pushing 80, was his usual curmudgeonly self. He didn't care a whit for small talk. Yet Sewell seemed to like being seated next to this church fellow from the country out West that produced the highly profitable pine furniture. But did Avery trust this young businessman who represented his church?

Montgomery Ward executives sat scattered throughout the tables of lesser suppliers. The mood was convivial, jovial. Avery, a teetotaler, made it easier for Charlie to decline the fine wine. They talked timber. Avery was born in 1873 to a wealthy Michigan family of lumber barons who turned to the pristine forests of the West as Eastern softwood declined. The Avery family owned great stands of Western forests.

As the evening progressed, Avery lost interest in the smooth speeches and began to fidget. Charlie noticed the industrialist scoot his chair back, bend down, and untie his shoes. Amused, Charlie watched, enjoying the old man's disinterest in decorum. But Avery wasn't finished. He removed his offending footwear and began rubbing his stockinged feet. The corporate scion, seated at the dais and soon to make a speech, didn't care. He placed a polished black Oxford on the table and leaned in to the young man from Harris Pine Mills. In a loud whisper he said, "Just between you and me, Charlie, don't ever buy Monkey Ward shoes. They'll kill your feet."

1. Clyde, age 8
2. James and Bertha, and the house James's war remuneration bought
3. Clyde, second from left, with pals on Milton's boardwalks
4. The Harris brothers, from left: Dette, Darrell, Jim, Claude, Clyde, George
5. Clyde's mother, Bertha, as a young woman
6. Clyde's father, James Harris
7. Clyde's grandfather, Stephen Harris
8. Clyde's mother in her later years
9. Mary Coe, age 9, and younger brother, Harold
10. Sam Kiester Coe and Susannah Frazier Coe: Mary's parents, married in 1873

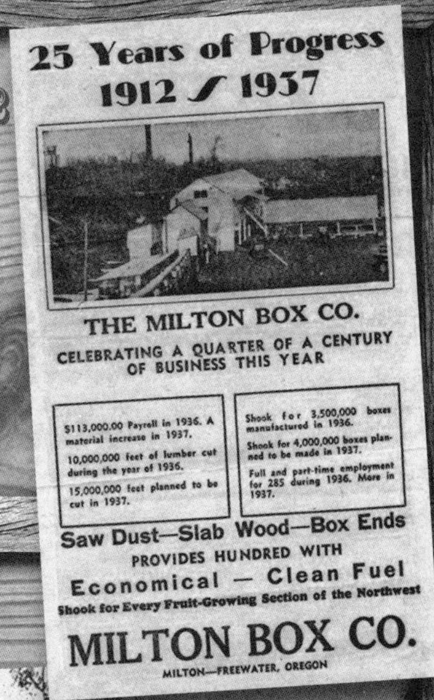

25 Years of Progress 1912 / 1937

THE MILTON BOX CO.

CELEBRATING A QUARTER OF A CENTURY OF BUSINESS THIS YEAR

$113,000.00 Payroll in 1936. A material increase in 1937.	Shook for 3,500,000 boxes manufactured in 1936.
10,000,000 feet of lumber cut during the year of 1936.	Shook for 4,000,000 boxes planned to be made in 1937.
15,000,000 feet planned to be cut in 1937.	Full and part-time employment for 285 during 1936. More in 1937.

Saw Dust—Slab Wood—Box Ends

PROVIDES HUNDRED WITH

Economical — Clean Fuel

Shook for Every Fruit-Growing Section of the Northwest

MILTON BOX CO.

MILTON—FREEWATER, OREGON

11. **Milton Box Company**
12. **Years of progress: Milton Box Company announcement in the Milton *Eagle***
13. **Mary, spring 1922, fishing on the Klickitat River**
14. **Harris boys Dette and Clyde and their wives**
15. **Tent living in the early days**
16. **The Winesap Road greenhouses, proof that Providence was at work even in the reverses that forced the company to move from town just before the Great Depression**

17. Clyde Harris, the young businessman

18. Mary Coe, new graduate ready for high school

19. Remnants of Blalock's old water flume for floating logs
20. Mary's mother, beloved Susannah Frazier Coe, affectionately called Aunt Susie by the community, died in 1946
21. Mary, about 94, beside a portrait of grandfather William Samuel Frazier
22. Dette and Marie Harris's twenty-fifth wedding anniversary, 1939
23. Mary, sister Lina Craig, and niece Margaret Craig deRomanett, taken in the 1970s

1. Army war posters
2. The war effort: making ammunition boxes
3. Bessie Freeman McGuire in 2012, age 100
4. Bessie's beautiful family in the 1930s

1

2

Random Reflections

BY ELSIE DICKSON

THE Old Lady in the Jet Bonnet is all a-twitter, thanks to clever pen and brush of Bob Moyer, artist, who mailed her a letter bearing, in lieu of an address, her picture. (See right.) But that, sakes alive, wasn't the half of it. Bob, who had read in last week's E. O. that the Old Lady is disappointed because the Clyde Harris hosses are coming from the South by train, when everyone knows they are WALKING horses, drew a most appropriate cartoon. And here it is, complete with the $10,000 gentleman walking horse and girl friends:

THE Old Lady in th
net is disappointe
that the Clyde Harr
are coming by train
nessee and Arkansas.
she said, "I thought t
WALKING horses!"

Showy Ho
Are Purch

Further proof—if any
that the Blue mounta
committed to light-type
superior breeding, and
way, is seen in the fact
Harris of Pendleton has
15 head of registered
walkers from recognized
in Kansas, Arkansas, Mi
Tennessee. The lot will
ed to Pendleton within
or so. They represent
ment of $30,000.

The collection, which
chose after infinite care
includes five mares, seve
colts, two geldings and o
The sire, named Midnight
Harris back a cold 10 gr
a female colt by Midnig
named Sun's Charm, and
months old, cost Harris $

Twice Grand Champ

All the mares which H
bringing west have been
Midnight Sun, who by the
national grand champion
therefore world champion
breed) in 1945 and 1946. Ha
tends no other horse of th
has duplicated Midnight
achievement.

One informed judge of ho
writes Harris that "these
bred, registered horses you
are just as good as the b
fords."

Relatively New Here

Not until last summer's

(CoCnt. Page 8, Col. 1, Se

Horses

Continued from Page 1, Sec. 2

Walla horse show did the aver
admirer of light types of ho
hereabouts have anything but
vaguest knowledge of Tennes
walkers. Several animals w
brought to Walla Walla coun
before the 1946 horse sh
competed in the classes s
n. The fans went for
characteristic of the

3

4

5

1. Clyde's prize, Golden Prince, with caretaker Frank Coe
2. Elsie Dickson opinion and cartoon
3. Clyde and Mary at Indian Village behind Roundup Stadium
4. Log trucks lineup that Clyde and Mary surveyed the afternoon of the board meeting, when the board affirmed the gift to the denomination. According to Richard Utt, author of *Harris and the Pines*, "Harris Pine Mills brings in as many as 50 truckloads of logs daily. The trucks are driven more than 250,000 miles a year."
5. Caterpillars

6. Clyde H. Harris and Mary V. Harris in 1953, when the gift was made public
7. Clyde and Mary in San Francisco for 1954 General Conference session, explaining their gift to the San Francisco *Examiner*
8. C. L. Torrey (left) and Charlie J. Nagele (CJN) on lumber stack beside Walt Applegate
9. Clyde and Mary in Hawaii, March 1954.

1. Aerial view of Harris Pine in its heyday
2. Harris Pine's old office building
3. Jim Wells, log scaler, on top of a load, right outside the old office
4. Charlie and Clyde in the 1950s, in front of old office
5. Unloading logs to cold decks, providing timber to cut in winter when logging stops
6. Unloading logs into a pond
7. Logs splash into the millpond
8. View of log pond and sawmill

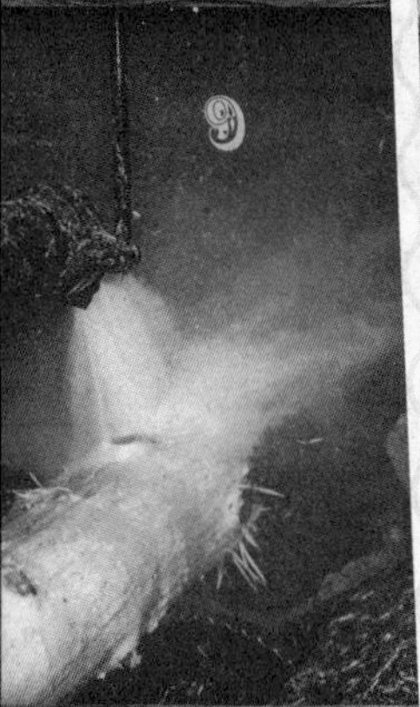

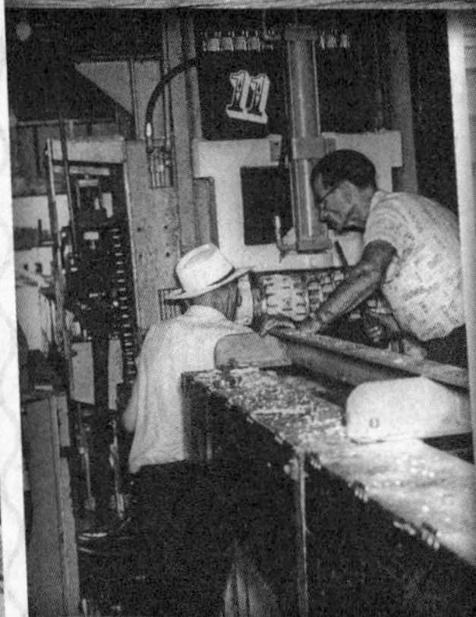

9. 1600-PSI debarker
10. Headrig grabs a log and takes it through the 16-foot band saw.
11. Tail sawyer hooks, flips, and straightens board to go down rollers through the sawmill.
12. Ed Paulson and Walt Applegate repairing Swedish gang saw, which slices smaller logs like a bread slicer
13. Sorting green chain; lumber sorted to size and grade
14. Straddle buggy taking a load from the green chain to yard or dry kiln
15. The forklift truck stacking loads in the yard, three tiers high
16. Harris Pine's new office building

1. Student Randy Phillips assembling furniture; he later became a pastor for the Upper Columbia Conference.
2. Another student assembling furniture
3. Shipping out the Harris Line (CJN and C. A. Scriven)
4. Bill Tapley with his children, Billy and Bernadine, on the log pond
5. Bobby Tapley, age 4, in front of the sawmill
6. Harris Pine's portable mill to be transported to Bill Tapley at Tollgate
7. See photos in Richard Utt's *Harris and the Pines* for more student labor pictures.

3

Firm 'Shoots' Business Stream With Fifty Thousand $2 Bills

FIFTY THOUSAND GREENBACKS—Looking over $100,000 in $2 bills that will be paid out by Harris Pine Mills this week are E. G. Johansen, left, HPM treasurer, and Rulon Smith, manager of Pendleton branch of U. S. National Bank of Portland. (EO photo)

By JIM EARDLEY
Staff Writer East Oregonian

Fifty thousand $2 bills began to pour into the local business stream this week as one of the county's largest employers sought to show the part it plays in the area's economy.

Harris Pine Mills began dispersing the $2 bills Tuesday noon. The process will be continued through noon Friday. Ninety per cent of the $100,000 is going to employes, the remainder to some suppliers.

About 75 suppliers who will be partially paid with the $2 bills have been asked to circulate them for the next four or five days.

Employes have been asked to use the payroll money to pay bills and purchase items rather than to immediately put them out of circulation by returning them to the bank.

THIS IS THE first time the company has tried the [illegible] petitive circumstances permitted.

The gimmick to show what economic impact Harris Pine has on the area's economy almost didn't get off the ground.

Rulon Smith, manager of the U. S. National Bank, which arranged the transaction, said the Federal Reserve Board in San Francisco was reticent to allow the idea.

A further complication was that President Kennedy recently announced there would be a recall of silver certificates.

BUT APPROVAL finally was granted and the 50,000 bills were shipped to Pendleton from San Francisco. Once here, it took two bank employes four hours each to count the bills.

Fortunately, noted Ray Stephenson, assistant manager in charge of operations at the bank, few of the bills were new.

Had they been, he said, it would have required more than eight hours' manpower because new bills stick together.

Traffic Gains on Three Of 4 Umatilla Highways

Traffic increased during October on three of four highways in Umatilla County, as compared with the same month a year ago, the state Highway Department [illegible] of 1961 indicates a 3.16 per cent increase when compared to the same period of 1960.

Ballots Go To Members

The names of 12 members today were submitted to Pendleton Chamber of Commerce firms in good standing, constituting the ballot for the 1962 election of board of director members, of which five are to be elected.

Chamber officials said members have until Dec. 12 to cast their ballots. On this date, the new directors will be named to the chamber's board. Those elected will each serve three-year terms.

Included on the ballot are Roy Comrie, Howard Hagen, Bill Halverson, Bill Hodgen, Carl Koons, Robert McCurdy, Robert McKellar, James Parry, Mike Sheldon, William Stram, Everette Stroble and Emerson Wheeler.

AN AMENDMENT to the chamber's by-laws in 1960 now makes it possible for member firms to cast ballots in accordance with a schedule based on the amount firms pay annually in membership subscriptions. Those paying from $40 to $74 as an example, are entitled to one vote and those paying $366 or more are entitled to six votes, the maximum allowable.

The ballots were mailed only to those members who are in good financial standing. It was reported that firms who have not yet paid and do so before the election date of Dec. 12 will receive ballots.

Shop in Pendleton
Friday Nights

1. 1960s administrative team and the brethren: from left, E. Johansen, H. Garcia, A. Stevens, E. Lorren, T. Lorren, C. Torrey, C. Scriven, unidentified member of the brethren, unidentified, E. Porter, M. Bain, W. Bain, unidentified, W. Wall Thorgeson, R. R. Figuhr, CJN
2. 1970s administrative team: from left, front row: Hollon Fry, Ray Watson, Charlie Fry, CJN, Thorgeson, Ernie Porter, Bill Hisey. Back row, third from left: Gib Garver
3. Newspaper photo, December 6, 1969: "Firm Shoots Business Stream With Fifty Thousand $2 Bills"
4. Montgomery Ward chairman Sewell Avery, whose loyalty to Harris Pine Mills was crucial, defying 1944 wartime order to unionize, being ushered out of his office by President Roosevelt's order. Time, May 8, 1944.

1. CJN and R. R. Figuhr in 1966 on porch of Clyde's cabin after Harris Pine Mills board meeting
2. A few of the Harris Pine Mill board members at an annual board meeting, Wallowa Lake
3. C. M. Bunker and C. A. Scriven
4. C. L. Torrey, treasurer, General Conference
5. CJN and N. C. Wilson, president of the North American Division
6. From left: R. R. Figuhr, Mary Harris, W. B. Ochs, Clyde Harris, unidentified, C. L. Torrey

7. Clyde with Senator Mark Hatfield of Oregon, CJN
8. Clyde and Robert Pierson, General Conference president, 1967
9. Glossy of *Listen* magazine, Clyde Harris
10. *Life* magazine, small picture of the Harrises
11. Two branch managers, Gib Garver and Jerry Calvert, with CJN

1. Clyde up to his chest in his wheat
2. Sign outside Clyde's favorite town, Joseph, Oregon (courtesy of Wallowa *Chieftain*)
3. The consummate fisherman
4. Intrepid woodsman
5. *East Oregonian* congratulations to C. H. Harris and feedlot
6. Feedlot, 1963
7. Cows munching pellets

8. Clyde's niece, Marcene Harris Garriot, 88, showing off a goat's beard, from a truffle mushroom treasure hunt
9. Young Charlie at his dad's sawmill
10. Willie White's youngest son, Francis White
11. Frank's motorcycle, with Charlie on back. He roared up a construction ramp into the administration building and into the dean's office, answering a summons to "see the dean."
12. Frank with his father Willie's car
13. Frank and Charlie at St. Helena, California, in 1930s
14. My parents, 1937, after my father turned down Spreckels Sugar's offer for the work at the Northern California Conference office as cashier, bookkeeper

1. **Harrises and Nageles admiring the new line of patio furniture**
2. **Kansas girl Lilly Hathaway and . . .**
3. **Little schoolboy Clyde**
4. **R. R. Figuhr jackhammering rocks from the logging road**

KIDS, CLASSES, AND SAWDUST

CHAPTER 24

In 1956, at my father's insistence, I applied for my first work permit. I was in no hurry to be off to the mill for a real job. Heretofore babysitting had been my biggest workload, other than Mother's tedious insistence that I was the designated floor waxer at our Ninth Street house, applying elbow grease on Aunt Mary's hardwood floors for the amber shine.

The last thing I wanted to do the summer of my sixteenth year was rise at the mournful sound of the HPM 6:00 a.m. whistle. But the day came when I forsook household duties and got up with the birds so I'd be in the furniture factory and ready for motion when the 7:00 a.m. whistle blew. I thought it a cruel way to start the first week of a summer vacation.

Just like the other student workers, I found myself bending over the low belt, and oh, my aching back! The job introduced me to muscles I didn't know I had and to the repetitive motions of gathering and stacking multiple-sized pieces of wood. Gradually I gained an eye for the boards assigned to me for any given hour. In partnership with other belt workers, I was instructed to gather and neatly stack on a dolly the correct size boards, as high as my head. Then I must quickly swing the dolly around and stack the far side. The trick was to turn the dolly quickly but not so fast as to cascade the first half of the stack across the floor—which was terribly embarrassing. For then the belt shut down while cutoff sawyers caught a break, and all the other low belt workers lounged beside their neat stacks. Catching boards, stacking boards, mopping sweat, drinking water, and downing salt pills—to my petted self this seemed like slave labor. Just the summer before I'd spent my afternoons trekking to the city pool, splashing it up with my pals, and slurping nickel Dairy Queens on the walk home as we danced on one leg and then the other to help shake water out of our ears.

The cutoff sawyers delighted in occasionally teasing or harassing a newcomer. That would be me! At a mutually understood hand signal all sawyers might saw just one size—the size I was to pick. A tidal wave of my size boards would flood the line, and I was helpless to do anything more than toss them to the cement floor, creating a pyramid of chaos. And that was their snickering point. I panicked and, of course, lost the battle. All the while the sawyers were peeking over their cubicles and cackling—knowing they were giving themselves a break as the alarm buzzer signaled that perfectly good boards were headed down the hog to be ground into sawdust.

By the next summer my upgraded assignment was tailing an edger in the furniture department. The methodical, overweight gentleman in charge of the machine had been at the same job for at least 100 years. His attention to detail and the scream of his saw cut out any small talk. I admired his steady, if slow, movements and the precision with which he measured and set the saw. As the cool, collected bosses strode through the factory (my father included), my jealousy roused. They had swamp-cooled offices and sit-down jobs. It wasn't fair. I was certain my eighth-grade graduation watch had stopped. The hours *must* be going faster than it showed. Across the aisle a handsome young guy labored over the sanding machine. I wished they'd assigned me to tailing there instead of with gramps. Even so, maybe he'd notice me if I put on a little more of the berry-stain lip gloss my mother had said was OK to wear.

Harris Pine branch plants were sprouting up around Adventist schools and hiring kids like me, who had other interests. Having a good time supplanted our work ethic. It would be years later—when my own offspring became work age—that I finally fully appreciated the value of Harris Pine's job opportunities and school tuition assistance.

Enter Gilbert Garver. He came to Pendleton long before the brethren did. And long before the church entered the picture, Clyde tapped him on the shoulder for something big, although at the time it seemed to Clyde to be just another step in manufacturing and growth.

As in so many of the stories that grew out of Harris Pine, Gib had no idea he would become an integral part of God's plan in denominational education. Now, he wouldn't take a fortune for all the practical knowledge he gained in working for the Lord nor for the wealth of understanding he had about his God, struggle though it was to develop the student work program. But first, his story.

In 1942 Gib needed a job, and he and his young wife were willing to endure the soot-and-smoke-belching train trip from Texas to Pendleton, Oregon, so he could get on at Harris Pine Mills. But Gib had barely acquainted himself with his assembly-line job when the mailman delivered his Greetings from Uncle Sam. He'd been drafted. Too soon he waved goodbye to his pregnant wife and was on his way to Pearl Harbor. When he refused to carry a weapon, he found himself carrying a medic bag over the bloodstained fields and towns of France. For the next five years he tended the grievously injured and the dead buddies that he often found facedown in the mud.

"I was sent to Europe and my brother went to the South Pacific," Gib reflects in his soft Carolina accent. "It was our mother who saved our lives with her prayers. Many times she couldn't sleep from worrying about our safety. One night she went out to her big garden, walked down the corn rows, knelt, and prayed. The Lord spoke to her in the darkness, 'Your boys will come home.' "

It was five years later when the barracks' radios blared news that the war was over. Gib had no idea where his brother was or what had happened to him. "But at the mustering-out processing center, Fort Lewis, Washington," he remembers, "I was walking down the sidewalk when out of the blue here was my brother coming toward me." With bear hugs of joy they traded stories of extremities and narrow escapes.

Gib says, "I told my brother, 'I just got a letter in the mail that I've got a second baby boy I've never met, and I'm going home!' "

"It took three days to get honorably discharged," he says. Finally, January 1946, he returned to Pendleton and a grand reunion with his wife and two small sons. Harris Pine's new ponderosa pine furniture had just begun rolling down the conveyor belt. Reggie DeVice, the man developing the new furniture for Clyde Harris, quickly made Gib the assembly line boss.

The young couple basked in the joy of resuming their lives together. But it was to be tragically short. A gas leak, a lighted match in the cellar as his wife shelved newly canned fruit, and an explosion! Six months after he'd swept his wife and two little boys into his arms, he stood, head bowed, his hand on her casket, his heart in shreds. The faith that had sustained him through the horrors of war now faced an unimaginable test. "I give the Lord all the credit," Gib says.

The people of Pendleton closed ranks around him during those terrible days. One Sabbath, just one month after the funeral, Mary Harris approached Gib. She had been watching his struggle with grief as he brought his sons, ages 2 and 5, to church each week. She touched his sleeve that Sabbath and said, "Gib, if you have any thoughts of putting your little boys up for adoption, Clyde and I would like to be considered first."

Gib was touched by the kindness and generosity of a couple he so admired. "They were just great people," he says. With that kind of support, he decided to stay in Pendleton and rear his boys, putting his whole trust in His Lord.

Gloria Poole was in her second year of secretarial science at Walla Walla College when she learned about the Garver tragedy from her parents and sister who worked at the mill. They lived in a Victorian home down the street from the church school. When Gloria came home that summer to work in the new furniture factory, she took an instant liking to the genial young foreman, feeling great sympathy for his tragedy.

Gloria remembers, "On Saturday nights, if no program was scheduled, we would skate in the gym. I was a good skater," she says. "Gib couldn't skate at all. I just wanted to help him learn. So I skated backwards around the floor and held him up."

Gib soon trusted her with his heart as well. Could she be persuaded to leave her carefree college days for marriage and instant motherhood? On a sultry July day in 1948, the gymnasium packed with friends, they promised to love and to cherish each other.

The couple were adjusting to marriage and Gloria's instant parenthood when personnel supervisor Bud Smith tapped Gib on the shoulder. "Clyde Harris wants to see you," he said.

"So October 1948, there I was sitting in front of Mr. Harris's desk, shaking," Gib recalls. But it turned out that the first furniture rolling off the assembly lines were such a success that HPM needed streamlined supply lines across the United States. Clyde had decided to build a plant back east. In his customary tempo, he had dropped into the metropolitan Washington, D.C., area midweek, rented an old abandoned brewery, and called Gib into his office the following Monday.

Clyde came right to the point. "Gib, we need better furniture distribution in the east. I'd like for you to move to Alexandria, Virginia, and build a branch plant for us. Last week I rented an old brewery back there. I want you to tear it out and build an assembly plant."

Gib says, "Gloria and I had been married just six months. I gathered my wits and told Mr. Harris I had to take it to the Lord first. Like Gideon, I wanted to pin it down to be to be sure." And herein lay Gib's strength. The Lord always came first.

"That night," he remembers, "I took a big ball of cotton and placed it in the middle of the lawn, which was already gathering dew. Early the next morning I went out, picked up the cotton, and took it inside. I thought it was dry as a bone. I asked Gloria to feel it. She put it to her cheek and said, 'It feels dry to me.'"

"When I got to work, I told Clyde, 'We'll go.'"

Alexandria, Virginia, was the first of Harris Pine expansions that would stretch to 26 assembly plants built during the next three decades. The young couple never dreamed to what lengths their faith would be tested. But they stepped out trusting the Lord.

"I didn't know that Alexandria, Virginia, was right beside Washington, D.C., and the cost of rent would be three times what it was in Pendleton," Gloria says. In fact, neither did Clyde. Nor was any adjustment made in Gib's small salary. He wondered if he had heard the Lord aright.

"We had to be on a railroad siding," Gib says by way of explaining Clyde's choice of real estate. "When the first carload of unassembled furniture parts arrived, I unloaded the boxcar myself."

In the next two and a half years, under Gib's supervision, the old brewery morphed into Harris Pine's first branch plant. The couple added a baby daughter and another son and struggled to pay their bills. Then in 1951, before the church took over, Clyde asked Gib to establish another plant in Chicago Heights. But that assignment wouldn't be for long.

"In 1952, as Clyde shook hands with the denomination," Gib reflects, "Charlie Nagele visited me and explained his vision for the student labor program. He asked me to move to Keene, Texas, and build a plant four miles from Southwestern Junior College. I was tickled to be going to Keene, for I had graduated from the academy there.

"Mr. Nagele made arrangements with the college to have a school bus shuttle the students in four-hour shifts." But dovetailing a work program with a college environment wasn't easy. He was the first to face the headaches of student labor problems.

In an anecdote Gib still finds amusing, he recounts a three-way conference call to the college president and himself, placed by Charlie Nagele. The problem was a headstrong college business manager dead set against working with the new plant manager and an all-hours college curriculum. Evidently the crisis demanded immediate attention, because the college president took the phone call while sitting on the platform during a chapel program. "The president understood my dilemma and gave his word that he would work toward solving the problem," Gib reflects, adding, "The business manager soon found other employment [*smiles*]."

The last thing on a teenager's mind, whether working for the summer or fitting work between classes and studying, was how to master gluing, nailing, stitching, and stacking. HPM always was world class in production, but it took many sacrifices and acceptance of inefficient ways over mechanization, often including less than optimum building sites in out-of-the-way school locations.

The headaches and faith testing continued as schools and workplace collided. In a letter dated June 18, 1964, and addressed to General Conference president Reuben R. Figuhr, my father enumerated the headaches of the student labor program. For instance, one of the challenges for the plant was that the Wage and Hour Division prohibited students under 18 near certain equipment such as power-and-air-driven screw and stapling guns. Therefore, while the plant's hands were often tied offering certain positions to younger students, employing 18-year-old students also meant that the plant had to (1) continually adjust to a high turnover rate (given that 18-year-old students are typically in their senior year) and (2) compete with the senior students' other interests.

"The above points out the urgent need of the respective educational institutions giving the industry they enjoy the more mature students. This is not now the case. School principals at the outset commit themselves to this in seeking the industry but soon lose their good intentions as their own interests mount and the students progress to senior status.

"I have a report on my desk now of two academies that have offered the 18-year-old students monitoring jobs for their senior year. These are our best and trained students and we must start all over again and replace them with younger (less than 18-year-old) ones.

"These senior students naturally prefer sitting in the dormitory corridors in the evening where they can study while being paid the minimum wage of $1.25 to working in our plant. Thus, without a proper school policy and their help, we cannot do much about it. We realize the school administrators do have problems also in surrendering their 18-year-old senior students, but it's simply a case of 'you can't have your cake and eat it, too.'"

"Absenteeism is also another problem. Every time a student doesn't show up, our assembly line is disrupted and shift changes take place every four hours with students.

"Here are some of the reasons, legitimate or not, that some of the principals accept for absence from work at their Harris Pine plant:

- Taking school pictures
- Examinations
- Makeup schoolwork
- School picnics
- Music festival
- Promotion tours
- Six-week leaves
- Sickness
- School discipline
- Free labor at school."

The issues and challenges of employing students at the plant and fostering some joint collaboration to iron out at least some of the wrinkles took time.

For 14 years Gib refined the work program in Cleburne, Texas. Eventually he moved his family to Braintree, Massachusetts, and built a plant for Pioneer Valley Academy.

His sixth and final plant was at Mount Pisgah Academy in North Carolina. Gib remembers, "One of Harris Pine's vice presidents, Tom Lorren, happened to see the round pine picnic table I had built for my own backyard patio."

Tom took the idea back to Pendleton, enthusiastic about building it out of nearly indestructible redwood, and the redwood division was born. Soon Gib had 16 girls sewing cushions for the popular redwood patio furniture.

"After I built the round pine picnic table," Gib remembers, "Charlie Nagele said, 'Gib, I wish I had a dozen guys just like you.'"

Hijinks happen when teenagers are involved. But Gib instituted a no-tolerance policy. When his own son lit a firecracker in the factory while on the job, he "made him walk seven miles home and laid him off for a couple of days." He hurries on to say, "But I want to brag just a little," as he recounts his son's dedication and successes in the Lord's work.

At each new plant Gib posted a sign over the doorway, words etched by his own handiwork into the pine board: "Do more than you get paid for, because one day you will get paid for more than what you do."

"I tried to instill those words into every boy and girl who worked for me," Gib says.

Like most missionaries, the Garvers are rich in memories. Until recently they enjoyed their retirement in the home Gib built for his family near the last branch plant he created, Mount Pisgah Academy in Candler, North Carolina. In 2013 Gib turned 92. His hair is white, but he retains the twinkle in his brown eyes as he sits beside his beloved Gloria, his wife of more than 65 years. They now live with their daughter.

He recalls with pride the letter Robert Pierson, then General Conference president, once wrote to him. It included this statement: "The work you are doing is just as important as [that of] the preacher in the pulpit."

I can attest that Harris Pine Mills was a sermon in work boots for me and a host of others.

SO THIS IS PENDLETON

CHAPTER 25

One thing the citizens of Pendleton can always count on is that tickets are half price on the second Wednesday of September, the day their much-anticipated Round-Up kicks off. It is Family Day, and all the schools let out at noon. At 1:00 p.m. the "Star-Spangled Banner" echoes across town. In the early days an American flag also shot into the sky above the stadium, and kids without tickets squinted into the sun to see which way the small parachute was floating as they raced against one another to capture the souvenir.

An online CNN travel blurb told it this way: "It happens each year for four days in the second full week of September . . . when this friendly, normally placid Eastern Oregon town of 16,600 at the foot of the Blue Mountains hosts more than 50,000 visitors for one of the biggest rodeos and premiere Wild West celebrations in the United States." The article tells the story behind the festival's slogan "Let 'er buck!" and the characteristically wild element of Pendleton's Wild West history. However, "these days," the article continues, "Pendleton is a comfortable, unpretentious, openly friendly, middle-class town that retains a Western flavor. While the Round-Up fiercely defends its traditions and Old West flavor, it too has become more family-friendly."[1]

Every year a large grassy area behind the arena is covered with doeskin teepees. It's the annual Native American exhibit that many liken to a family reunion. During the years I was growing up, Native American women *and* men wore their hair in long braids and came to town wrapped in colorful Pendleton blankets and laced-up doeskin moccasins. Now during the festivities, Native American ponies and their riders assemble for the colorful ceremonial ride around the arena. There are bucks, princesses,

[1] Travel, "Oregon: Lines Form Early for Venerable Rodeo Classic," *CNN.com*, Feb. 13, 2004, accessed Jan. 18, 2006, www.cnn.com/2004/TRAVEL/DESTINATIONS/02/13/pendleton.rodeo.ap.

and chiefs bedazzled in feathers, beaded finery, and white doeskin. The Native Americans allow the curious to bend down and peer into teepees, getting a glimpse of how life perhaps used to be. Visitors stroll across the grass, stop to stare at babies strapped to cradleboards, and take photos of tenacious pint-sized bucks practicing their dance steps for the center arena finale.

In years past paper boys and flag chasers, including the one I married, gathered under the grandstands to collect empty pop bottles for spare change, working against the hazards of the careless crowd above. They quickly learned the certain whistle of an empty Orange Crush dropped from above, and those quick on their feet avoided a bonk on the head. Alerted by the crowd's roar when a writhing bull unseated a favorite cowboy, the kids hoisted themselves up and peered around the legs of spectators to catch the action free of charge.

Every year Mary and Clyde enjoyed an arena box of 10 seats, midcenter, close enough to make one's heart lurch if a runaway Brahma bull tried to leap the fence. Invited guests from across the country were treated to a close-up look at the Old West. Native American cultural skirts, Levi's and dinner-plate-sized belt buckles, Western shirts, and cowboy boots ruled the day, and Mary delighted in wearing her Western finery astride her Tennessee Walker horses in the Dress-up and Westward Ho parades.

Juxtaposed against the rough and ready afternoon Round-Up was the evening show, Happy Canyon. There a colorful pantomime of Wild West history unfolded against the night sky. By the time a lovely maiden plaintively trilled a traditional Native American love song across the open air arena with a brave who took a bullet in the heart and plunged over the painted plywood cliff into the water below, loud music and whiffs of liquor escalated from the rowdy hall behind the stage. For some a "few more" at Goldie's Bar climaxed the evening with harmless drunks trying to weave their way home. The celebrations finally wound down Saturday night.

Pendleton's beginnings date back to the first wagon trains trailing up and over the Blue Mountains and their sheer, precipitous shale. It was considered the most hazardous part of their journey. After bumping and roping wagons down these sheer walls, upon reaching the valley below, many unnerved immigrants decided this was as far West as their courage would take them.

But the Cayuse, Nez Perce, and Umatilla tribes were already here. With the White man's sense of entitlement swiping huge swaths of land from the Native Americans, it was inevitable the cultures would eventually collide.

Unfortunately, the settlers sought to gain the favor of the Native Americans by giving away free whiskey, smokes, and trinkets. And "holy" men were not far behind, bringing a black Book that supposedly pronounced the "Indian" ways of dressing and rampaging just wrong. Indeed, the Native American peoples made a strong effort to embrace the White man's God, who was, sadly, often represented as a stern, unyielding deity.

But with the endless clouds of dust that signaled the wagons streaming into the country came diseases such as measles and smallpox. Whole tribes were powerless to withstand these mysterious ailments and almost certain death. A few among the tribes revolted at the mixed messages of the White man and his stern God who seemed to bring them certain death. When chaos, rampaging, and murder broke out, for a time the encroaching civilization stopped in its tracks.

But an irrepressible wild town sprang up at the foot of the Blue Mountains, taking its name from the 1864 U.S. Democratic vice presidential candidate George Hunt Pendleton, who, with his running mate, lost to Abraham Lincoln and his vice president.

A curious part of the area's growth was built underneath the board walkways. Oppressed Chinese laborers—the men who laid the transcontinental railroad track and thus stitched the country together—tunneled underneath the town to escape racism and abuse. Alongside underground saloons, they carved out their own society to survive. Mountain men and cowboys came to town for baths, shaves, and laundry. The baths got cheaper as the day waned, and the water murkier, until the last baths were the bargain of the day, a dime. Everyone knew that bootleg whiskey, gambling, and a chance to clean up was underground. But topside, in the heart of the city, the 32 saloons and 18 bordellos weren't ignored either.

Pendleton's history is best summed up in the story of Til Taylor, the upright legendary lawman who served as Umatilla County sheriff from 1902 to 1920. Pendleton won its reputation on toughness, hard riding, hard drinking, and taking the law into its own hands. By the early 1900s the growing region's county seat was in Umatilla City, a once prosperous region about 40 miles north, on the Columbia River. But with its fortunes on the decline, Pendleton had big ideas. Moses Goodwin, an early pioneer who traded a span of horses for 160 acres from a squatter, and his influential friends wanted the courthouse in Pendleton across the street from his log cabin hotel. He lobbied his judicial friends to change the county seat to Pendleton. Things drifted along for a while until he got a judge to order that the records be brought to Pendleton from Umatilla. Umatilla balked and stalled.

One quiet Sunday morning (some say it was before dawn), a band of men broke into the Umatilla Courthouse and triumphantly took the records to Pendleton. The booty was lodged in a makeshift shed Goodwin had conveniently provided, and Pendleton was proclaimed the new county seat. Citizens of Umatilla were furious. They sued. Nothing came of the suit, and Pendleton has been the Umatilla county seat ever since.

From 1920 to 1933, the prohibition era, moonshine was illegal, but card games and gambling were allowed. The speakeasy was born. Warned to speak easy, so as not to raise attention, and not to tell, patrons were allowed entrance. An outside lookout posted on a boardwalk bench would signal a police raid. At the sight of anyone who looked like a lawman the bench sitter pulled a cord that jangled a series of bells. Seconds later all evidence of hard spirits and shot glasses had disappeared, leaving a few earnest card players with no evidence of anything to quench their thirst.

Every bar downstairs had a gold scale, and when gold miners came with their pouches of gold dust the bartenders were ready. A pinch of gold dust on the scale, and a drink was served. Tipping wasn't in vogue, but bartenders constantly swept the counters with towels to catch the residue of gold dust. At the end of the shift the floor was swept and mopped, the bartenders threw out the dirty wash water, and stomped their boots in the mud, making sure they were wet and muddy. Then they came back in to walk back and forth behind the bar. Their shoes caked with mud and gold dust were later panned out in their room—the tips for the night.

In spite of such shenanigans, the turn-of-the-century lawman Til

Taylor is remembered fondly as one who lived by a code of principles and honor. On the east side of town a 12-foot statue of horse and rider towers over a lily pond and reflecting pool known as Til Taylor Park. For decades kids playing tag around the large stone base would occasionally gaze up at the kindly-faced lawman, holster at his side, astride his ample bronze horse.

Til gained his reputation with his uncanny ability to track criminals through the impossible underbrush and deep ravines of the Blue Mountain foothills. He spent days in the wilderness in hide-and-seek chases and gun duels. But it was his policy to capture but never kill an outlaw. For 22 years Taylor called Umatilla County back to law and order, but on a sultry July afternoon in 1920, that changed.

The attack was so sudden—just as cells were being unlocked and lunches handed out—that the five prisoners overtook Taylor, and his gun fell out of his holster. One prisoner shouted, "Shoot 'im!" but Til hit the gun barrel and diverted the shot. It was hand to hand from then on.

Again the shout "Shoot 'im!" and the gun fired directly into Til Taylor's heart.

"Guy, I'm shot," Taylor gasped to his deputy as he slumped to the floor.

The fugitives escaped to the railroad tracks and jumped onto an unscheduled freight train rumbling east into the Blue Mountains.

The cry "Til's been shot!" rang out, and the community roused like never before. Even the lawbreakers agreed that Til Taylor was an honorable man. Anyone harming him would have to answer to all the people of Umatilla County. Men in saloons and men lounging under shade trees leaped to their feet. Hardware store owners unlocked their doors and pointed to gun racks and ammunition, nodding, "Help yourselves." Bloodhounds were brought in from Walla Walla Penitentiary.

By popular demand, Til's brother, "Jinks" Taylor, a prominent area farmer, was pressed into duty to serve out his dead brother's term as sheriff. For six days 100 men combed the dense brush and forests, and the five desperadoes were finally rounded up. Two were found sleeping in an old sheepherder's tent near La Grande, 50 miles across the Blue Mountains from Pendleton.

When word spread that the killers had been caught, the courthouse grounds swarmed with angry men seeking justice for Til. They surged against the building, threatening to break into the jail and mete out justice themselves.

In what looked like a scene right out of an old Western movie, the ensuing events were reported:

"Then above them in the open door of the courthouse appeared a figure with bared head and in shirt sleeves. It was the newly appointed sheriff [Til's brother "Jinks"]. Beside him stood a guard with carbine in hand.

" 'Boys, if Til were alive,' the gleam of the streetlamps glistened, reflected in the moist glisten in his eyes, 'he would want you to let the law take its course. You who are friends of Til, I ask you to do as he would wish if he were alive. Rest assured justice will be done.' "

"For a moment there was silence, then from here and there sounds, words of approval came from the compact group. Calmer citizens supplemented the sheriff's appeal.

"Little Pendleton upheld its record for intelligent action, common sense and upholding the law. Slowly, haltingly, still in a state of partial indecision, the crowd turned its back on the jail and trailed to their homes. As the sheriff had promised, law took its course and justice was done."[2]

As Pendleton lurched into the twentieth century, development, not the least of which was Harris Pine Mills, buffed up the town. Yet remnants of the Wild West tenaciously hung around. In 1940 the tiny Adventist church on SE Eighth Street, adjacent to the Umatilla River, with no running water or indoor plumbing, grew from a membership of 30 to 300. As Seventh-day Adventist families spilled into the ravine and surrounding hills, they liked to believe they had something to do with the more "civilized" Pendleton. With the Harrises' substantial help and the donation of four lots of land and a house from a local citizen, a larger church was built at the staggering cost (by 1941 standards) of $25,000. The free house became the church school with teacher Louise Beers and five little boys. Almost at once the school grew to 60 students.

Mary Harris entered the sanctuary on the first Sabbath morning in their newly built church, surveyed the packed congregation, and whispered to Clyde, "We've built too small." But the Harrises were practical people, and work began immediately on a gymnasium. One Friday before the roof was on, when Clyde saw that a storm was rolling in, he told sawmill superintendent Wes Henderson to stop his work at the mill and bring

[2] Charles Wellington Furlong, *Let 'er Buck: A Story of the Passing of the Old West* (New York: Overlook Press, 2007), pp. 56, 57.

his crew to the gym to finish the roof before sundown. The resulting gymnasium featured rolled roofing, and it inevitably became the perfect place for children to play Annie Annie Over—throwing a ball back and forth over the high, rounded roof. If the ball didn't go over, a shouted "Pigtail" resounded to the other side. In the fall of 1948 church services moved to the Quonset hut gymnasium, and the school took over the church building.

It was in the mid-1940s that two 10-year-old Adventist schoolboys, one of whom I would marry, sold newspapers downtown after school. Each afternoon they'd sprint to Main Street to gather their stacks of *East Oregonians* to sell for a nickel each. As Rodney and Larry prowled the streets, they were alert to surreptitious doorways and steps that might add clues to the tales they had heard of underground tunnels, secret passageways, and mysterious goings-on beneath their feet.

Indeed, their curiosity was whetted as they followed inlaid glass sidewalk squares that allowed light into the caverns beneath, and one day the boys rode an open freight elevator to the underground labyrinth of tunnels. As they crept along a dirt corridor, they saw rooms to the side, bare lightbulbs hanging from ceilings, bunks built into a dank corner, and Chinese men lying down smoking long pipelike apparatuses. This was excitement no Uncle Arthur's story could rival! Then they discovered dust-laden steps leading up.

They were stopped by two doors at the top of the stairs. One couldn't be budged, but the other suddenly gave way, spilling them into daylight, a gaudy bedroom, and a gaggle of shocked creamy-skinned women who stared speechless at the two young scoundrels. The befuddled schoolboys were way over their heads. Then peals of laughter, whistles, and giggles burst forth from the women as the two kids banged, skidded, bumped, and fell down the stairs disturbing years of dust in a cloud so thick they could barely see to make their escape. Scared witless and sure the police would come looking to arrest them, they promised each other not to tell a soul.

But at least one of the boys told his tale over the supper table that night. His dad, Walter, sternly shook his finger at his boy and said, "Never try that again." The boys kept to the sidewalks from then on.

The small pink-and-yellow neon sign advertising "cozy rooms" in the upstairs window of the Medemach Building had always been there. Stella Darby ran a business on Main Street that would make most of us blush, should we blunder in some cold night wanting to rent a room. But Stella ran her establishment with an elegant grace that one does not expect when we read about the seamy side of the good old days. Besides her own comfortable and elaborate upstairs ambiance, her sitting room, drawing room, she treated her girls equally well, providing immaculate quarters. Stella had a pious side, and installed an upstairs chapel with pews as a place of worship for her girls, with an itinerate parson preaching each Sunday. Today all these rooms are open, and for a fee Underground Tours of Pendleton will escort you on a guided tour through the tunnels and up the staircase.

I learned that Stella was firmly in place upstairs when Harris Pine Mills moved to Pendleton in 1939. She married Uncle Johnny, a soldier boy she and the girls had patched up and nursed back to health and to whom the girls looked to as their benefactor and protector. He announced forthcoming raids, and both the patrons and the girls would hustle for the escape routes to the underground.

One Sunday in 1953, as news of the gift Clyde and Mary Harris had given to the Adventists spread across the nation, the local Presbyterian minister threatened to read a list of Stella's patrons in church. Suddenly, it seemed, Pendleton got religion. The coincidental timing couldn't have been more welcome to the brethren. But Stella merely plastered over her Main Street front door, moved around the block, stayed in business, and kept paying her taxes. She retired in 1965, moved to Walla Walla, started receiving Social Security, and when she died at the age of 76, the attendance at her funeral topped any on record.

OUR BACK DOOR

CHAPTER 26

When we moved to Pendleton, my parents had every intention of choosing a suitable, affordable house, just not in Clyde's back pocket. At first my parents bought a Sherwood Heights bungalow in an addition east of town, just up the hill from Stover's Market. With only 800 square feet of living space, my mother loved it; her housekeeping duties were simple. We weren't in a hurry to move again, but Clyde made it clear to my father that our present home should be temporary.

So with Clyde's nudging, my folks began reducing the possibilities. On North Hill (as opposed to South Hill's Harris Heights) was a modest, newly built house. My parents liked both its clean, spare lines and the price. It was tucked into the lower regions of North Hill. (The hilltop was reserved for attorneys, bankers, and retired wheat ranchers.)

But the problem with that location became abundantly clear when my father casually mentioned his interest in the North Hill bungalow to Clyde. He was met with a long silence. Apparently, to live on North Hill would be a traitorous act. So that part of town was eliminated. North Hill was Clyde's competition. South Hill was his domain.

Dr. deRomanett, niece Margaret's husband, a man with soft humor and a twinkle in his eye, was heard to parody the subtle tension between the two regions. North Hill, rising prestigious and proud, had the elegant Mormon church gracing its summit. Rising high on the south hilltop a new Seventh-day Adventist church would reside (but not until 1958) on an opposing promontory. With both hills presiding over the town's valley, each choir taking their hymnbooks and preparing to sing—with great sincerity the Adventists sang, "Will there be any stars, any stars in my crown?" and the Mormons, holding their hymnals aloft with firm and equal devotion, sent back the words of that grand old hymn "No, not one."

Of course, in actuality the town enjoyed a friendliness and balance that didn't betray any biases but added a spark of energy to our lives.

Our location choices boiled down to one: 615 SW Ninth. This was the Harrises' old home, designed by Mary and hastily built with Clyde's oversight and construction crew in 1939. The house stood on an imposing corner lot, jutting above a sturdy rock wall much like the stone embankment of the their later and permanent Goodwin Avenue residence. The only problem was that it wasn't for sale.

Even so, Clyde arranged a tour of the house for my parents. I suspect he coaxed the congenial lady of the house into selling, probably pointing out she was in an excellent position with a buyer (my parents) just waiting for her to say yes.

So one afternoon we toured Clyde and Mary's old house, entering through the back door, which was conveniently just beyond the sidewalk. (Clyde didn't waste space on yards.) The door opened to a dark, steep staircase that dropped to the basement. We stared into the cavernous subterranean space and the garage beyond with its low ceiling, owing to the sunken living room. My father would have to remember to duck as he came through from the underground garage or be clobbered by the hefty ceiling rafters.

As soon as we stepped inside the house, we were enveloped in a sensory overload that put our good Adventist biases on alert—the attending color, haze, and tenacious odor of Camel cigarette tan.

One thing modern decorators stress is that if a house is small, carry one color continuously throughout. And Clyde and Mary's old house met those specifications perfectly. Except for a wide black header painted around the kitchen wall—it was Camel tan throughout. I could tell the friendly chain-smoking woman liked novels as much as she enjoyed smoking, for hundreds of Western paperbacks of the Zane Grey's *Riders of the Purple Sage* genre ruled the shelf space.

A quick inspection of the closets showed that Clyde had thought better of wasting square feet on closets. A cubbyhole with a few hooks to hang work clothes in each of the two bedrooms sufficed. Had Aunt Mary agreed to this design? Where did she hang her treasured fur coat and the pretty dresses she wore to church?

The owner was willing to sell—for just above the asking price of the North Hill bungalow. My parents agreed. The Ninth Street house would be our new home. Clyde was delighted. Now we were part of his paternal oversight. My father was to be under his tutelage, why not the whole family? He had a vested interest in seeing Charlie succeed, and that included seeing his family happy. And what better way than to pay close attention to us? The Harrises could tell when our lights went on in the morning and off at bedtime. Clyde could spot my father's car sliding to the house at the back door for an occasional lunch break.

In the winter the living room was my favorite. In one corner a sizable black floor grate delivered deliciously warm air from the basement behemoth, an oil furnace. The air billowed my circle skirts and burned crosshatch patterns on the soles of my saddle shoes. It was also quirky and voracious. Every six months the oil delivery man hauled his hose to the hookup pipe and unloaded fuel for the basement beast. At times, for reasons unknown, the furnace turned dyspeptic. We'd hear a bang and belch from the basement, and out of the grate would spew a cascade of smoke followed by a fine black oily haze that misted quietly over the furniture. Mother was not amused.

But there were some lovely benefits to living there, not the least of which were the oriental cherry trees ringing the lower front yard's two confluent streets. In March and April the buds burst open into a forest of glorious, delicate pink blooms.

Once they said yes to the house, my parents worried about the starkly apparent needs—fresh paint throughout to cover the R. J. Reynolds atmosphere and carpentry needed in the kitchen and closets. They learned another side of Clyde that unexpectedly emerged soon after they signed the legal documents to own the house.

A few weeks before we were scheduled to move in, my mother and I went to the mill to pick up my father. No one thought of owning two cars in those days. When she knocked on his office door, my mother discovered Clyde sitting in his old office chair and my father with his hand on the doorknob. Clyde motioned for my parents to sit down for a moment.

What followed was an uncharacteristic display of Clyde's feelings. First, he expressed gratitude for their "leap of faith" in accepting the call to Pendleton and for their willingness to buy the house he suggested. "Paint the place," he told them. "Replace the yellowed floor linoleum, resurface the hardwood floors, and add some closets." He handed my father a check for $1,500. "But don't tell Mary."

That comment, my parents learned, was not a malicious attempt at cutting Mary out of his dealings but to protect her from what she would deem his "unnecessary expenditures." To Clyde, she was still the child bride who worried they were really poor. She never outgrew the fears of those first fragile years of learning thrift as Clyde plunged into expanding the Milton Box Company. Success would come. She would agree to the gift, but she never felt secure.

My father leaned heavily against the office door, his knees unreliable, his eyes moist. He had sensed this side of Clyde when he took on the burdens of Harris Pine Mills. But until now Clyde had maintained a gracious but distant presence. By this expression, it was clear that he was an astute discerner of the needs of others.

That welcome windfall only happened once during my parents' Harris Pine Mills years. But Clyde knew when to tangibly say that he understood. It boosted them immeasurably. He gave at the time it meant the most.

The Harrises worked hard during the day and visited friends in the evening. They often enjoyed the easy stroll from their house to our back door. Our supper was usually in progress or being cleared away when we'd hear the back door open and their cheery hellos as they stepped into the kitchen from the landing. My mother worried over our family's household disarray in contrast to the stark cleanliness of their home. But to them our house felt like home.

A Christmas tradition was quickly established after we moved to town. On Christmas Eve Clyde and Mary enjoyed a quiet dinner and festivities at our house, bringing a small gift each for my brother and me. Christmas Day, beginning at dawn, they celebrated with their beloved neighbors Dr. and Mrs. Easton and their four lively youngsters, with knee-deep holiday frenzy.

We also lived a block from the steep embankment and wooden steps Clyde had built for easy access over a cliff to the church and Quonset hut gymnasium at the bottom of South Hill. Every Friday afternoon the deacons transformed the gymnasium into a church, setting a thousand folding chairs in neat rows. By then the steepled church was a 10-grade school. An Adventist parsonage, built by Clyde, perched on the brow of the hill so close to the steps that each morning I could see what the pastor's family was having for breakfast.

I loved our proximity to school and church. No more waiting for a school bus on rain-swept corners as I had done in Portland. From slamming our back door to breathlessly sliding into my school desk, if I sailed down the steps three at a time, I could make the sprint in five minutes. Ice and snow doled out a few pratfalls.

The hike up the steps after classes was a case of wishing away my after-school chores and piano practice. I remember the clothesline at the side of the house, the white sheets flapping, so white and crisp and fresh. The black soot flecks from the HPM wigwam burner, the bane of fussy housewives, didn't make the clothing any less fresh. We never worried about it. After several years of irate letters to the editor in the *East Oregonian* complaining about the soot problem, HPM countered twice over time to remind the region of Harris Pine's contribution—once paying wages in silver dollars, and a second time with two-dollar bills. Both times the entire county was flooded with the awkward currency followed by a quarter-page letter in the *East Oregonian* from HPM president Nagele explaining the mill's ongoing process to clean the air and reminding the community of the economic boom HPM brought to the area. The mayor ran up the white flag with a newspaper reply. Enough already! We like sooty sheets.

In any event, Harris Pine finally eliminated the insult to perfection by pouring thousands of dollars into its clean air project.

Waiting for me every Friday when I got home from school was a large white ball of flour in the kitchen sink. Mother made it my job to wash away the flour in order to leave the protein, the gluten. Dad loved her "featherless giblets." But her famous Sabbath dinner roast—the spineless mass fed through the meat grinder with onions, nuts, eggs, seasonings, and bread crumbs—was her crowning gluten event.

The Harrises now owned a new full pedal Consonanta organ that Clyde had ostensibly bought for Mary. In reality I think it brought back memories of his ragtime days and he was eager to revive the natural skills that had earned him extra money playing in the dance halls. He introduced the musical instrument to us with a rousing rendition of "Turkey in the Straw," his joyful guffaws revealing how much gold he had in his teeth.

The instrument graced their front room wall opposite the Mason & Hamlin grand piano. Aunt Mary said that if I took lessons, I could practice every afternoon on the organ. So with an offer like that, my parents added organ lessons to their kids' enrichment budget. For lack of sufficient space in the living room, Clyde had seen fit to put the two massive speakers down the hall across from the master bedroom, essentially at the opposite end of the house.

My organ career was never stellar, and unfortunately these practice sessions took place during Aunt Mary's nap time. As I worked the bass pedals, sometimes my feet landed in the right vicinity but not exactly. Much of the time I played blithely on repeating the musical faux pas, Aunt Mary getting the full amperage as she napped. The speakers boomed the sour notes to her long before they hit my ears.

During those afternoon interludes in my ordinary life, tiptoeing through the whisper-quiet house, so lovely in every detail, I came to know Aunt Mary. Periodically she suffered with a bad back, usually brought on by a fanatical cleaning spree. When that happened, she might be curled up in excruciating pain for days, as pain medicine didn't seem to help. She simply suffered, lying in her king-size bed, the room dim, the blue silk coverlet pulled around her face, her body contorted in pain. My heart went out to her in the sad, luxurious atmosphere.

When I finished practicing, she'd call out for me to come and sit on her bed and visit. She'd take my hand, and I'd chatter away with news beyond her four walls, hoping to bring some cheer to her dark bedroom.

I remember running home one afternoon to comb through my favorite books, picking a stack I was sure would lift her spirits and make her smile again. I remember giving her a rousing pep talk about naturalist Sam Campbell's book *On Wings of Cheer*, the story of a red-winged blackbird Campbell and his wife, Ginny, had nursed back to health. As I warmed to the story, giving her the details, I realized how much she was enjoying my animated report. I don't think she ever read the book. It was the visit she craved.

She loved gossip. It diverted her attention to the joys and trials of other families. Babies were her obsession. She didn't talk about her childlessness, but she was always up to date on the current crop of babies in the church and her extended family. She'd tell me about holding a little bundle, and her eyes would shine with tears.

Clyde bought the first TV in the neighborhood. With a tall roof antenna aimed imprecisely toward Portland, Clyde boasted that his was a color TV, and indeed it was color. Various shades of chartreuse shimmered with static and snow, and no one had the courage to ask Clyde why all the other colors were strangely absent.

He was the channel master, lining up rows of straight-backed chairs in the tiny den off their living room. He expected company a night or so a week and called to remind us when Sergeant Bilko in *You'll Never Get Rich* or *The Sixty-Four-Thousand-Dollar Question* was coming on.

Romance was Mary's darling topic. With an abundance of nieces and nephews, she kept a fat scrapbook of wedding invitations and newspaper pictures of family and friends. Matchmaking topped her list of favorite pastimes. I vividly remember participating in one.

MARY THE MATCHMAKER

CHAPTER 27

Now that we lived in a small town and didn't travel the camp meeting circuit every summer, and despite the fact that the yard behind Harris's old house had no fence, I theorized that owning a dog would be no trouble at all. In the first harried months after our move to Pendleton my parents were noncommittal. Yet one frigid January Sunday just before our move to Harris Heights, my father suddenly appeared at our Sherwood Heights home and thrust a soaking wet mongrel puppy into my arms. A sympathetic log roller had scooped her out of the log pond where she'd fallen, and my father, who happened to stride by at the moment, was handed the bedraggled little creature. I loved her the moment I saw her. But distemper quickly followed her cold midwinter swim, and despite my constant ministrations spooning nourishment, she died.

But my few brief days with her fueled my ardor for a dog. When the Sears catalog briefly offered puppies for sale, I wore the book out. But realizing that the Harrises didn't care for pets and had none of their own, my parents decided we shouldn't have any either. The matter stayed there for some months. I prayed fervently anyway and took in a stray shepherd that appeared one night. I was certain it was a direct answer to my prayers. I remember my mother's apology to Mary about the dog. She told Mary that he just appeared one night and I believed it was a miracle. But soon after taking his fill of my specially prepared meals, the dog ambled off, probably to gladden another girl's heart.

Rather than pets, the Harrises loved visitors, especially during the Pendleton Round-Up. Clyde had Harris Pine purchase box seats for personal and business friends, and few could resist an invitation to travel to the Wild West for the adventure. A favorite guest any time of the year, but especially during Round-Up, was George Yeoman, Jr., Harris Pine's top company salesman based in Chicago. Mary had a particular fondness for the young, irrepressible George of Yeoman and Son. The older Yeoman having retired, it fell to George to carry on his father's tradition with Harris Pine, always cutting the monetary deals in a way that suited Clyde. George was a regular visitor to Pendleton.

Shortly after my family moved to Pendleton, George Yeoman was widowed. With the untimely death of his young wife, he found himself coping with loneliness and two little sons to rear alone. Aunt Mary grieved and fretted, treating him more like a favorite son than favorite salesman, and George was fond of Clyde and Mary, too. He had been a true gentleman when the mill changed hands. His loyalty to Harris Pine never wavered. He respected Clyde's religion, and if Clyde was giving the mill to his church, George supported it.

Mary hatched a plan—the perfect solution to George's unattached status—and she proposed it to Clyde. They would introduce George to Mary's single niece by taking them both on a camping trip. Clyde, with a surprising lack of judgment, agreed to the plan. The niece was a young and beautiful flight attendant, never married but very desirable in the days when the rigid rules of air travel ruled out all but flawless women. Mary reasoned that she'd supply the opportunity for romance, solve George's loneliness, and in the payoff George would become family.

George was born in Chicago and grew up in Chicago. He had never been camping in his life. His father was Clyde's first marketer and instrumental in bringing Avery of Montgomery Ward and Harris Pine together in a lucrative collaboration, as well as Sears. With both companies headquartered in the windy city that prided itself on the biggest furniture market in the nation, George played a crucial role in the Harris Line of Super Pine.

Mary's plan, or plot, as it turned out, would include our family. Our job was to make sure the guests were transported, fed, and entertained with sufficient chitchat to avoid awkward moments for the unwitting couple. Clyde and Mary would host a camping trip to the brim of Hells Canyon, a mile-and-a-half-deep gorge, the deepest in North America. There Mary would introduce her niece to the round, amiable Midwesterner, George, and his two sons, ages 4 and 7.

The trip began with the long drive to Wallowa Lake, three hours from Pendleton, and heading east from the tiny burg of Joseph, Oregon. The highway wound 30 miles into a canyon alongside the Imnaha River, past Big Sheep and Little Sheep creeks, Clyde's favorite fishing grounds. There the pavement ended, and the trip then took on an adventure that city-bred George was totally unprepared to appreciate. Ignoring the signs recommending four-wheel vehicles only, Clyde assaulted the steep one-lane dirt ascent to the heights above the Hells Canyon with his little HPM pickup. In those days it was no trip, or camping, for sissies, as the summit campground perched on the canyon rim. Clyde and Mary visualized it a fitting spot for George and Pretty Niece to meet.

The forest fire lookout tower loomed above the one-and-a-half-mile-deep gorge, providing another 82 feet above terra firma from which to view the dizzying drop into the yawning chasm and, at the bottom, the winding sliver of the mighty Snake River. The breathtaking scenery, it was hoped, would evoke romance.

So far, so good. At least it seemed to Mary who must have dreamed up the spirited excursion in the middle of the night. Clyde passed the news on to my father, who sprang it on my mother. Suddenly we were part of an unusual camping trip, by way of Wallowa Lake (Clyde's cabin), then dirt road for four torturous hours recommended for off-road vehicles only. Of course, we didn't have a four-wheel-drive transportation, and neither did Clyde and Mary. Our old Packard would have to do, winding up and up to the campground at the very end of the road. There we could gaze at the mysterious mountain range, The Seven Devils, when we were not watching our footing or peering far, far down at a tiny ribbon of water glistening at the bottom.

Food and bedding were my mother's department, although Mary was never a slouch when it came to providing food for any occasion. As the plan developed, George and his two sons would fly into Pendleton on a beautiful July day—and quickly be spirited on to Wallowa Lake. As with many a grand plan—it looked fine on paper, but expediting it was a bit tricky. You see, Mary and Clyde had overlooked a few details.

George was a city boy through and through. He had been to Pendleton several times, but his only experience with mountains was reading *National Geographic*. He had yet to experience a trip to the mountains on Clyde's terms. Couple that with a surprise introduction to a bright, beautiful flight attendant on the brink of Hells Canyon (to say nothing of the trip getting there), and with George's phobia of heights, would he be in any shape to countenance romance?

Clyde contributed to the camping trip by hauling the mattress out of their cabin and plopping it into the bed of his pickup, declaring it "for Mary." Mary toted enough food for a week, although the adventure was just for one night, OK, two nights if things went well and sparks were ignited. I don't remember what my parents contributed, although I know we arrived at the top of the steep chasm with sleeping bags and food galore. We co-opted a picnic table.

Then Clyde and George arrived. As soon as he escaped the pickup, George dropped onto the table bench, his face pale, saying that he'd never been so terrified in his life. He was still trying to digest being subjected to four hours of single-lane dirt road with a terrifyingly yawning chasm on one side. His little boys, however, were loving it.

Aunt Mary and Pretty Niece had made the trip up the mountain in Mary's Cadillac, and most of the day was gone by the time they reached the camp site. I remember how pencil thin and beautiful she was, dressed fit to kill, manicured nails, not a hair out of place. So with the lookout tower even closer to the canyon rim than the campsite, Clyde declared we all should climb to the top to see the "view." George blanched even whiter. He had been gazing down at "the view" all day, straight down, holding on to the pickup's bouncing seat for all he was worth. But Clyde wouldn't allow George to demur.

The lookout tower arose high on triangular stilts with a series of ladder steps leading to a trapdoor 82 feet into the sky and the glassed-in lookout room. I think George was behind me. I remember that I shared George's fear, because as I neared the top rung, I glanced down (not a good idea), and my father, who was already in the room above, saw my bloodless face and quickly hauled me into the room with both arms.

He didn't have to grab George. Halfway up the ladder was view enough for George, and he gingerly backed down to terra firma. By this time post-traumatic stress disorder had kicked in big time and he was hardly up to small talk. But while Mary clucked brightly around the campfire, he tried to make nice with Pretty Niece. Mercifully, we all climbed into our sleeping bags early.

George didn't sleep that night. Granted, wilderness blackness intensifies the brilliance of stars, which appear as though they are standing on stems, waiting to be plucked. But the glorious sight was completely lost on the city boy who was in a state of hypervigilance and mental and physical exhaustion. By contrast, his sons slept like kids on a Boy Scout trip.

The rest of the story I heard secondhand from my father. George quietly woke him before dawn. Would he take him and his boys down the mountain a little ahead of the camping party, i.e., the Harrises and Pretty Niece? He simply couldn't face any more adventure and matchmaking. He was feeling ill. He thought he might be catching something.

So my father and George carried his little boys to the getaway car. (How all that didn't disturb the rest of the camp remains a mystery.) A short note left on the picnic table provided an explanation, sort of. With the car in neutral and no headlights, they coasted out of earshot, escaping George's worst nightmare. Charlie assured George that he'd convey his thanks and appreciation later. Two tuckered-out little boys didn't protest.

At the foot of the three-hour descent (it was faster going down) they reached the settlement of Imnaha, which boasted a post office and a cracker barrel store. Up until recently the town had hosted an annual Bear and Rattlesnake Feed. As George finally saw a semblance to civilization, he began feeling better and better. At the sight of a real outdoor pay phone hanging sideways on the weathered store siding, he asked Charlie to stop. He bounded up the creaky steps, past the life-sized wooden statue of a Native American person and the twig-and-bark rocking chairs, and grabbed the phone receiver. The get-the-sales-contract George was back and on a mission. Nothing would deter him.

Charlie heard one side of the person-to-person call—from Imnaha, Oregon, to Chicago, Illinois—collect! At the other end of the line a schoolteacher—whom George had been dating—tentatively accepted the charge. She had grown tired of waiting for him to make up his mind, and though she probably was tempted to not accept the call, she did, and George made it worth her while. At 6:00 a.m., Imnaha time, surrounded by an immense forest of ponderosa pine and Douglas fir, and under the stolid gaze of the wooden statue, George proposed to Janet. The patient woman heard the certainty in George's voice. He was absolutely clear.

By Thanksgiving Clyde and Mary and niece had forgiven George. He and Janet were married. The boys had a new mother and . . . they all lived happily ever after. Really!

Of course, in the after telling, we all had a wonderful adventure and enjoyed the fun of Mary's matchmaking. But wait! George didn't forget his harrowing mountain escape from which he had felt sure he would never return.

Just before Christmas George placed another call, this time from Chicago to Harris Pine and my father.

"Charlie," George boomed, "I want you to take your family to the airport on Christmas Day. I'm sending Marilyn and Chucky something that must be picked up promptly."

I think he told Dad more than that, but Dad remained mysterious. Christmas morning we were in the car on our way to the very same airport where the flyboys of World War II had trained years before. This time a DC-3 landed right on time, a couple of people disembarked, then someone lifted down a barking crate. I say barking because the noise and the bouncing crate portended a new level of hijinks at our house.

We gathered around as Chucky unlatched the door. Out bounced a 6-month-old wire fox terrier pup, full of fizz, the answer to my prayers! I don't know how George knew about the embargo on pets at our house. At any rate, George took care of the problem. After all, if George sent her, she'd get to stay. Nevermind that she terrorized the neighbor's chicken coop, tore a hole in the postman's pants, made a sprinkler out of Clyde's watering hose, and howled under the street light at midnight if we didn't make sure she was safely in the basement. Oh, and chased off all of Clyde's pet chipmunks at his cabin. But Checkers stayed because George sent her. If that isn't a terrific thank-you gift, I don't what is.

WILD BILL

CHAPTER 28

Remember Paul Bunyan, the mythical lumberjack of tall tales and giant height? The legend records it took three storks to carry him to his parents. As the yarn goes, he and his companion, Babe the blue ox, with an ax dragging behind, dug the Grand Canyon. The rocks he piled to douse his campfire now stands as Mount Hood.

It would be hard to top Bunyan legends. But Harris Pine Mills lore harks to one of its own. That would be Bill Tapley. He had a small waist, arms as big as stove pipes, and hips and legs like cannon barrels. Kids who read the Sunday *Oregonian* comics noticed Bill's close resemblance to Popeye's nemesis, Brutus. But no one said it out loud.

Wild Bill, as he was known around the sawmill, had a big presence and an even bigger mouth. When a machine acted up, he'd turn the air blue, but he could fix most anything with the screwdriver he kept in his back pocket. It almost got him killed one day.

Bill grew up in South Dakota, where the winters are long, and in those days farmers were poor. Bill and his seven rowdy siblings didn't know the word *etiquette*. One day during breakfast one of Bill's brothers threw a fork across the table to shut him up. Now, it's understandable that Bill's colorful mouth might rile an older sibling, especially when being called names by a kid brother. Well, the fork stuck in Bill's eye. Blood spurted. His sisters held their heads and screamed. The commotion escalated as Bill ducked under the table. They were sure he had lost an eye, but Bill wasn't about to admit he was wounded. Yanking the fork out of his eye, he popped up and singsonged, "Na-na-na-na, you didn't hurt me!" The fork just missed the cornea. Perhaps the incident set the tone for Bill's life. Accidents trailed him like a swarm of hornets.

In those days there was no such thing as a Department of Social Services, and truant officers didn't show up when Bill decided he'd had enough of school after the sixth grade. So when Bill left home at age 13, nobody made a fuss. Poor boys did things like that back in 1915 and Bill got a job herding sheep. Such a job can be pretty tough on a 13-year-old, and the night's mighty lonely in the wilderness. When winter set in, Bill learned to sleep wet. Some mornings he had to pry ice off his overalls to get his legs apart. For 17 years Bill lived the lonely life of a sheepherder. Once his saddle horse rolled on him, but the saddle horn protected him from suffering more than deep bruises. Rumor had it that one day a prairie rattlesnake lay coiled right where he stepped, and it struck and fastened onto his leg. Bill crawled into his bedroll and slept it off. But, as the story goes, the snake needed first aid. One logger said, "Bill was like a grasshopper. If you pulled a leg off, he'd grow another one."

In the divine scheme of things God saw a tender heart inside that unrefined exterior. Bill was 30 years old when he came west to visit his sister who lived in Coeur d'Alene, Idaho. And one memorable Saturday night Bill met a special woman at a party. Adna Meeker, a Seventh-day Adventist graduate of Yakima Valley Academy, was tall and quiet with a refinement and poise that even rowdy Bill couldn't ruffle. When he proposed marriage, Adna said no. Bill wasn't deterred when she added, "You have to become an Adventist first." He didn't think that would be a problem if Adna came with the deal. He knew nothing about the Bible or God. Like everything else he tackled, his response was "Just lead me to it."

He didn't see it coming, but God crept in. "He gave his heart to the Lord," his son Billy says. "My dad told me he smoked his last cigarette on the way to his baptism. He threw the butt out the window into the Coeur d'Alene River and never smoked again." That was in 1934. The couple married that same year.

When they moved to Pendleton and Harris Pine Mills in 1945, Bill was already missing two fingers. The way Bill went at things, if he didn't know math he couldn't have counted very far. His first Harris Pine job was sorting and piling lumber on the green chain. Piling green lumber all day suited him, but he had one little problem. When he got riled over a balky board or machine, no matter who was around, he took the Lord's name in vain. It earned him a reputation around the mill.

The old-timers remember the night he was shoveling snow on the steep metal roof over the green chain. Someone told him he ought to tie a rope around himself in case his boots slipped, so he just looped the rope around his hand. Of course, his boots took off like a toboggan—and the rest of Bill followed along. In a heartbeat his 240 pounds were dangling by his hand in midair, and so did his unrepeatable cry for help.

A forklift driver came along, but getting Bill down was a problem because a crowd had gathered to see him hanging by one arm off the green chain roof and swearing like a sailor. The loud guffaws drowned out Bill's protests, and rescuers had to stifle themselves while they caucused on how to liberate him. He truly was hurt. Dr. Easton set several bones in his arm and told him to behave himself and practice safety from then on.

When Bill was promoted to sawmill nightshift foreman, he sometimes took his two little boys, Billy and Bobby, and sometimes little sister Bernadine with him to work until the 8:00 p.m. lunch break. Billy remembers watching the logs winched up into the sawmill from the log pond. "One night the guy running the sawmill log lift was slow, and it irritated my dad," Billy says. "One log was halfway onto the deck, and my dad kicked the mammoth tree. It rolled off the platform and hit the water 15 feet below with a humongous splash." That was pretty exciting stuff for a 9-year-old.

Bill roamed the sawmill and green chain, often spotting and fixing trouble on the fly. One night he was up on the green chain machinery repairing it when he reached down and grabbed one of the bumpers to steady himself. (On the green chain the boards travel down a slot, hit the bumper, and spill into a bin.) A 2' x 12' Douglas fir ripped down the slot and banged the bumper *and* Bill's finger. The digit went flying. Bill swore, wrapped his handkerchief around the stub, finished what he was doing, then walked to the first-aid department.

Then there was the day Bill split several of the 2' x 12' steps leading into the sawmill. The carriage set works, the weighty machine that set the width of every slab, needed to be taken out for repair. It was so heavy the men were afraid to lower it down for fear it might slip and fall into the log pond. Bill studied the problem and said, "Take it apart in two pieces. I'll carry 'em down."

Like a world-class weight lifter, he grunted as he pressed the biggest piece slowly to his chest. The crew held their breath. Bill's legs bowed and buckled. Outside, the top step creaked and began to split until his foot rested on the next step. It too bowed and cracked, as did all the others on his way down. Then he brought the smaller piece down and dumped both parts into a pickup that sagged so low it needed more air in the tires. The guys cheered and slapped Bill on the back.

"Aw, it was nuthin," he grinned. The story got loose and spread to the box factory and the office. Everyone wanted to see the steps that Wild Bill had buckled. His legend was well on its way.

If something happened to the 16-foot band saw whining through the logs on two large steel wheels, a lot of energy had to dissipate before it was safe anywhere near it. If it ran into embedded rocks or metal, some of the teeth would sheer off and the flying bits of steel penetrated like lethal bullets. On rare and dangerous occasions, the saw would crack. Experienced mill hands knew to duck, clear out, and flip the power off.

One day Bill heard the 16-foot band saw making a clicking sound, the first warning it had a crack and might disintegrate into lethal shards if not immediately shut down. But Bill didn't want to shut down the entire sawmill to investigate the cause of the clicking. Instead, he carefully pressed the end of his screwdriver against the flat side of the saw.

The saw exploded like a World War II battlefield, splitting into flying shrapnel. Pieces of hot steel shot everywhere. The force knocked Bill flat on the floor against the large steam cylinder called the shotgun that powered the carriage. Pieces of the band saw flew like bullets, taking out the floor, the roof, and the two-foot square pillar that held the saw in place. It destroyed the whole area. About 20 feet away Don Hathaway, the resaw operator, grabbed his neck as hot steel tore his flesh.

Someone got the power flipped off. For a moment nothing moved. Where was Bill? Two guys made their way through the destruction to see if he was alive. They found him pinned down flat on his back looking through the new hole in the roof, cussing and trying to pry his leg off the shotgun steam cylinder. With serious burns and a mangled hand, Bill was gingerly escorted to the hospital. Young Billy remembers looking at the bare bone in his father's hand. Dr. Easton told Bill to go home and stay off work until his wounds healed. Of course, he was back at work the next day. At Bill's next appointment Dr. Easton saw the dirty bandages and knew Bill had defied his order. He wasn't happy. "I told you to lay off work. You're to stay home until the bandages come off," he told him.

But Bill fashioned an elaborate bandage over the doctor's handiwork and went back to work. Before his next appointment he took off the dirty bandages and smugly admired the still-clean ones.

Dr. Easton smelled a rat. "Did you go to work?"

Bill grinned. "You told me not to."

It took two weeks to rebuild the sawmill, floors, walls, and roof. Walt Applegate stressed the point that he could stand in the center of the building and watch the birds fly over.

There was the time Bill was spelling off the edger operator, pushing a board through that didn't need sawing. It was coming through without a sound, so Bill raised his hand to signal it was coming; otherwise, the board might catch the guy like a guillotine. As Bill waved for the guy to get out of the way a large knot zinged off the top of the edger. At bullet speed it sheared off another of Bill's fingers. Blood spurted all over the boards. He wrapped his handkerchief around the rest of his finger and kept working. That afternoon as Lilly Hathaway was running some errands downtown, she saw Bill. His finger dripping blood, he was muttering as he strode toward Dr. Easton's Main Street office. Adna was scurrying behind him.

In later years Bill operated the Tollgate gang mill in the Blue Mountains. The gang saw cut small logs into lumber the way a bread slicer sliced bread. Bill made a good boss, but sometimes the fellows got under his skin. A theology major joined the Tollgate crew one summer, and he remembers one day when the guys lined up for chow. Bill wanted to say the blessing, but music blared from a transistor radio, and the men were laughing and talking over its noise. As the man remembers it, "Bill lost his temper but got the job done by bellowing, 'Turn the blankety-blank radio down so we can pray!' "

Bill was always sorry after losing his temper. He once blew his stack in front of Charlie Nagele. Charlie loved Bill, and Billy remembers that Mr. Nagele gave his dad a talking-to and Bill cried. But when Bill lost his temper, everyone felt it.

"I knew it bothered him a lot," says Billy. "He would talk to us kids about how wrong it was."

Bill had only a sixth-grade education, but he loved to study the Bible, and his favorite book was *The Desire of Ages*. He taught a Sabbath school class for 20 years. His class always reported the highest offerings and frequency of daily lesson study for the Pendleton church. Everyone loved Bill and admired his straightforward honesty. His Christian graces grew over the years. In his final illness, as the days of his life waned, he took great comfort knowing that he would awaken to see the face of Jesus. Adna would say that he became the most patient man she had ever known.

There's a biblical character that closely resembles Wild Bill. His name is "Rolling Stone," or, put a little differently, "Loose Canon." Peter—the unpolished fisherman with the colorful mouth. In fact, after three years of close contact with Jesus, and after putting his promise of fidelity on the line, Peter turned coward and, to save his own neck, cursed a blue streak. As the cusswords flew from his mouth, his forsaken Savior turned toward him, and their eyes met. They both knew Peter's betrayal. Peter fled into the night, and, as the Gospel of Luke says, he "cried and cried and cried" (Luke 22:62, Message).

Peter, great preacher, friend of Jesus, meet Bill, great Sabbath school teacher, able deacon, and friend of Jesus, both made soft and compassionate by their Lord. Someday, I believe, they will be instant friends. Their common bond, the endlessly seeking Savior, the "Hound of Heaven," who never gives up. "Ah, fondest, blindest, weakest, I am He Whom thou seekest!*

* Francis Thompson, "The Hound of Heaven," *Bartleby.com*, www.bartleby.com/236/239.html.

THE LAST FORESTER

CHAPTER 29

"As long as I can pass the yearly requirements for the red card, which is to do three miles with a 45-pound pack on my back in 45 minutes, I'm going to keep fighting fires," Dale Jenner says. "I spent 19 years out back in the woods for Harris Pine and as many with my own business doing the same things."

Jenner handled the forestry division of Harris Pine Mills the last 19 years of the company's existence. He saw the company through the bankruptcy, became friends with the court trustee, John Mitchell, and declared, "The company was liquidated with as much reclaimed dignity as was possible, given the circumstances."

Dale grew up in Pendleton. His parents met in Granger, Washington, as did many of the couples who early found Harris Pine Mills and Pendleton the place to settle down and raise a family, thanks to the steady jobs Clyde Harris offered.

Dale attended the little Adventist grade school, delivered newspapers after school, aimed the *East Oregonian* over the rock-wall embankment to Clyde and Mary's front porch, and every month, though in awe of the couple who lived in the grand house on the hill, knocked on their door and collected his payment.

In his family's tradition of hard work, his father from Tennessee, his mother from Colorado, meeting and marrying in Granger, Washington, they found themselves at Harris Pine Mills just when Clyde gave the mill to the church. Like the other families connected to the mill, Dale's parents wondered if the company would continue to thrive. And then, in spite of the hoopla, things steadied.

It was June 1970 when 16-year-old Dale started work on the low belt at $1.25 an hour. Two summers later, when he was about to be promoted to superintending the belt plus the 18 mischievous sawyers, Dale declined. He said he wanted a job outdoors. The bosses happily complied, and he found himself soaking up the 100-degree rays while pulling weeds in Harris Pine's back lot. But the job had a bonus. He got to drive a truck between the factory and the Pilot Rock retail store 11 miles south of Pendleton. He loved it. His bosses, Henry Garcia and Bessie Freeman McGuire, who managed the Pilot Rock retail store, made certain that he kept her store supplied.

It wasn't all roses. Garcia was a tough and demanding boss, but the freedom of the fresh air and the open road reinforced his long-held desire to become a forester. Dale admits without bragging that he did a good job for Henry, despite the day he didn't properly tie down a load of lime and swung his rig too wide, rounding Albertson's Grocery. The full cargo dumped across the intersection. At first he panicked. But on this memorable occasion his boss paid him the ultimate compliment. He joined Dale in shoveling up the powdery mess.

After Gem State Academy Dale obtained a college degree in forestry, then married his academy sweetheart, Judy. All those years Dale remained focused on his goal and returned to Harris Pine as assistant to forester Martin Ladd. When Ladd retired, Dale was promoted to HPM forester.

Dale's idea of punishment would be going to an office job every morning. By his own admission he'd rather be slashing brush in subzero weather any day than cooped up in a building. In fact, these days Dale is up and out of his house before dawn, often headed up I-84, commonly called Cabbage Hill, one of the most hazardous stretches of highway along the interstate, toward his government contracts to clean out brush and undergrowth in the forested areas.

Dale loves tough assignments. At Harris Pine he could estimate a stand of timber on the fly, how many board feet of wood in a given stand. Then he'd go to timber auctions with firm figures in his head on how to bid, and his buys kept the guys in the front office happy.

Dale knows all about Blalock Mountain. During his years as forester he personally kept keys to the gate that stood at the bottom of the eight-mile grade. He likes to reminisce back to Clyde's hunting days. The elk were protected on the steep forested terrain, and with Clyde's cabin at the top it was his private reserve.

"I had a lot of coaching from Charlie Nagele," Dale remembers. "Charlie had never been a forester, but his training under Clyde Harris, his knowledge of the timber, and the fact that he knew that a blue-stained log is unsatisfactory led to some stern lessons. Temperature and lack of moisture, such as cut logs left to lie too long on the ground before being loaded onto trucks, turn blue."

Dale recalls those early years when "Clyde used Charlie as his bird dog for his elk hunting." Every fall hunting season Clyde and friends mysteriously disappeared for long weekends. Inevitably, in the early years, my father was an essential part of Clyde's hunting party. Even though he never had a taste for hunting, he knew that if he was to be Clyde's understudy, part of the job description was being Clyde's bird dog.

Charlie's predawn assignment included hiking down Blalock following the terrain, flushing out the sleeping bucks that scrambled for higher ground. Clyde was ready for them, perched on a high knoll, aiming his rifle. Sometimes the shot wasn't so clean and easy. If Clyde's best shots felled the animal partway down the canyon, it was Charlie's job to skin and pack the carcass to the top and into Clyde's pickup for delivery to the meat processing plant. Clyde always came home rejuvenated and rejoicing. My father came home bone-weary.

"Clyde had a method," Jenner says. "He wanted Charlie to learn the terrain, the draws, and the size of the timber. What better way to teach Charlie to pace the mountain, learning the forest, than to traverse the wilderness while scaring up venison for Clyde."

To Clyde's pioneer ways, security meant enough meat for the winter. If his freezer was full, he was happy. As for Charlie, his reward was wilderness training at its best and a couple of butcher-paper-wrapped parcels from the meat packer, compliments of Clyde.

"Buttons," Clyde and Mary called them, a favorite accompaniment to a Harris breakfast—elk wieners, sliced and fried.

My mother's face grew red with embarrassment when people at church asked her where Elder Nagele was on fall weekends. She thought hunting and the work were mutually exclusive. But my father left the discomfort to my mother while he hiked and talked the Sabbath away with Clyde atop Blalock. And before dawn on Sunday morning he staggered out, half asleep, to conquer the mountains and learn about the timber business into which he had been so abruptly thrust. Clyde's boot camp!

Dale reminisces about Ard Stevens, Harris Pine's first forester, a bear of a man whom Charlie hired in 1957. Indeed, much of the lore about Charlie's bird dog days comes from those years when Harris Pine was developing a more solid commitment to tree farming, clean logging, and forest replanting. Ard joined Harris Pine as overseer of the growing empire of timber, improving the untenable habits of loggers who slashed the wilderness into heaps and piles. He started timber farms, planting new trees with an eye to the future. He also kept a constant vigil against summer forest fires, which were a constant concern to Harris Pine and the forest service.

Forest fires are the bane of the wilderness. Lookout towers, usually tall, spindly-laddered sky dwellings, are scattered through the Northwest timber ranges. They are usually built on mountain promontories with a 360-degree view and are staffed by an intrepid college student looking for a monastic summer adventure. At the first sight of smoke winding skyward, the lookout person radios a report of the suspicious spiral, pinpoints it on the map, and radios the location.

A few hardy couples, married right after Walla Walla College graduation, might head for a wilderness lookout . . . for their honeymoon. High on a mountain with just each other and boredom interspersed with teeth-rattling peril, togetherness takes on a new meaning. Although safe from grizzlies and cougars, a chance-black cloud appears and the idyllic scene changes in an instant.

One lookout was scraping old gray paint off the refrigerator generator. He was 65 feet up in a tower perched on the rim of the mountain, leaning over the top of the refrigerator. With a crick in his spine he straightened for just a moment. *Crack*! A seismic explosion like an M-80 firecracker ricocheted around the 12' x 14' room. It blistered his dog's nose and blew out the propane motor. On an otherwise-sunny day a bolt of lightning, out of a black cloud directly overhead, blew out three Forest Service radios and his new $500 ham rig, and melted into a puddle two inches of solid copper antenna on the roof.[1]

When it seems as if the weather can't get any hotter, suddenly nature can go still. The horizon, so cloudless, quickly gathers a ring of black clouds. As the gloom marches on, a hot wind begins to blow. In the distance, a growling or roar might be indistinct, but to the initiated it spells trouble—thunder. Animals are the first to sense the danger and instinctively search for cover. True to the warnings we humans usually ignore, the sky darkens above us, and big splats of rain make mud pies on our just-washed cars,

---

[1] Ray Kresek, "The Little Black Cloud," in *Go Tell It on the Mountain*, ed. Jackie Johnson Maughan (Mechanicsburg, Pa.: Stackpole Books, 1996), pp. 11, 12.

progressing until the wind, the thunderclaps, lightning, and rain envelopes us in a summer storm. Firefighters and smoke jumpers recognize the jagged lightning slashes as a warning that if a tree receives the jolt, it will begin to blaze. A dry forest is no match for nature, especially if there is an abundance of undergrowth and forest management hasn't cleared away the offending brush in a while. A forest firestorm is in the making.

This is where Chicken Neck reappears as recorded in General Conference president Robert Pierson's 1974 devotional book—a fire encounter that defies logic.[2]

[2] Robert H. Pierson, *Faith Triumphant* (Hagerstown, Md.: Review and Herald Pub. Assn., 1974).

THE LORD'S CAT

CHAPTER 30

The call came from the Forest Service in the Heppner area, 40 miles northwest of Pendleton. "Lightning strike. Can we borrow an HPM Cat and skinner?"

Log boss Ard Stevens didn't hesitate. He loaded up one of the big Cats and delegated the expert, Chicken Neck Jones, for the job.

"It's a dry, grassy area and threatening timber and the town," the ranger told Ard. The conflagration was sweeping across eastern Oregon, threatening Heppner. All the available Oregon Forest Service rigs had been deployed. They needed more. Of course, Harris Pine would send in their best workhorse of the logging industry, an $80,000 Caterpillar that built logging roads for Clyde and moved mountains of earth, rocks, and trees.

Jones knew the drill. He'd rev the blade against trees two feet through, shoving and rocking until they toppled, creating a clearing to diffuse the oncoming blaze. But only the best cat skinners had the experience and judgment in timing to back away at the right moment as the emerging stump popped from the ground. An unskilled skinner might misjudge his timing, upending his machine and his ill-fated self.

The Heppner forest fire had already caused a lot of damage. From the forest came the sound of an eerie treetop whooshing that any skilled woodsmen would recognize as indicating that the fire had gone from dangerous to deadly. Every living thing was in danger.

The fire had turned fickle. Jones and his Cat faced an unconquerable foe. Columns of black smoke lay heavy over the area. The blaze was jumping the hastily plowed clearings feverish firefighters had hacked among the trees and fast encroaching on a prime stand of timber.

Through binoculars the rangers could see the fire skip through the tops of the pines, rushing toward the ground crew.

"Pull back! Pull back!" forest headquarters radioed its units. But Jones was already surrounded. The inferno was fast leaving him and his Cat no escape.

Chicken Neck hadn't earned the respect of Clyde Harris and loggers by being a quitter. In fact, if Jones could be faulted at all, it would be for not cutting and runing sooner. With his escape route all but ablaze, he leaped off the Caterpillar. Running for his life, he flew across burning grass, dashed through the blazing underbrush. His feet felt as if they were on fire, and the stench of burning rubber from his boots filled his head. When he finally got to safety, his clothes were sooty and burned, his shoes were charred, and the pads of his feet were blistered. But he was alive.

Ard Stevens was the first to learn the details. He felt personally responsible for sending his man and machine into harm's way. He had risked Chicken Neck's life. Dispirited, he radioed the Forest Service with the message "We lost one good Caterpillar, but saved our man." Harris Pine was self-insured, and Ard knew the company would have to absorb the loss of the $80,000 piece of equipment. But he took comfort in Chicken Neck having escaped.

Two days later the fire spent itself at the side of the highway some miles south of the spared town. Ard toured the burned-over area to see the damage for himself. Blackened treetops told of the inferno that had skipped through trees, racing capriciously, licking at limbs and logs as it roared toward Jones like a freight train.

In hopes of finding the charred rig, Ard tromped through the devastation, around burned tree hulks, over charcoal logs. Then he stopped and listened to a strange buzz emanating from the debris of the forest. His boots covered with ash, he kept walking, certain that he heard a motor running somewhere. He was in the worst of the damage now. He saw Forest Service equipment "charred, twisted, melted, by the intense heat of the fire—that is, all but the Harris Pine Mills 'cat'! The huge earthmover stood defiantly in the midst of the black disaster—in a circle of greenery—its motor still running. It was untouched by the inferno that had destroyed almost everything around it. The foliage in the tops of the trees above the 'cat' was black and lifeless. But the 'cat' purred on—a miracle of God's protecting hand!"*

Ard sprinted toward the purring rig. It still had 50 gallons of diesel fuel in its tank. He paused a moment at the incredible sight. Then he hopped on, shoved the gears forward, and drove the Cat to the highway to take it home.

---

*R. H. Pierson, *Faith Triumphant*, p. 181.

666 LAKESHORE DRIVE

CHAPTER 31

"Certainly you can remember that number . . ."

My father smiled as he faced the audience, giving a report on Harris Pine Mills at the 1964 General Conference session in San Francisco.

"We invite you to visit our national sales headquarters and permanent showroom at 666 Lakeshore Drive, Chicago, Illinois. . . . This is a beautiful showroom of more than 4,000 feet of space, and we are proud of it. Here you can see all the products manufactured by our furniture division. We are known favorably and well among buyers everywhere. They refer to the owners of Harris Pine Mills as 'the church' and sincerely do so with respect for it.

"Most manufacturers traditionally give whiskey to large buyers during the holiday season. If on occasion we [give] anything at all, it is Oregon apples. We have had many respond that they appreciated them so much by comparison. I am sure our competitors would do well to switch to apples, as they continue to have a hard time before the Harrises' fast competitive pace at the marketplace."

At the turn of the century commerce had eliminated the gentle afternoon carriage rides wealthy folk enjoyed along meandering Lakeshore Drive at the edge of Lake Michigan. Buying and selling transformed the lane to a pounding thoroughfare through Chicago's business district. Now modern rush-hour traffic clogs the wide streets, and the beautiful lake and appreciable location hardly get a glance as consumers and achievers focus on destination rather than ambiance.

Early-twentieth-century merchants grasped the energy of Chicago, the windy city. Twice yearly furniture makers and retailers came to Chicago to scout new products and designs in furniture. Chicago became a hub of commercial buying and selling. In 1934 *Time* magazine observed:

"Over the entrance of the 30-story American Furniture Mart Building in Chicago's Lake Shore Drive is a bas-relief of a woodman hacking a tree into logs, a sawyer cutting the logs into lumber, a carpenter fashioning the lumber into furniture. Under this symbolic device last week hurried thousands of furniture buyers from big stores and throughout the land to attend the twenty-first semi-annual exhibition of the American Furniture Mart. Elderly, gray-thatched Wade McGowin, head buyer of Wanamaker's, went from Manhattan, as did tall, dark-haired Mike Joseph of Gimbel Bros., and Charles S. Shaughnessy of R. H. Macy. From Sterling & Welch in Cleveland went short, heavyset George Killius, known as one of the keenest buyers in the trade. In the largest collection of new furniture under one roof in the world, they, and some 6,000 other buyers, joked and jostled over Mickey Mouse chairs, Biedermeier boudoirs, imitation Louis beds, functional sofas, juvenile hat racks, free wheeling beer carts. . . .

"The Chicago mart is not the only furniture exhibit but it is the most important. . . . There the big retail stores select suites (pronounced "suits" by most of the trade) for display in the autumn, when the public does most of its furniture shopping. The professional buyers who last week smiled, frowned, scratched their heads, slapped feather mattresses and dusted armchairs with the seat of their pants may not actually close a deal for weeks or even months. But the mart claims 70 percent of all furniture purchased in the U.S. is bought from companies exhibiting at its winter and summer shows. . . .

"Biggest exhibitor at the Chicago mart, Kroehler Manufacturing, claims the distinction of being the world's largest maker of upholstered furniture. . . . Customers today include such heavy buyers as Sears, Roebuck, and Montgomery Ward."[1]

Harris furniture didn't take up residence in the halls of commerce until late in 1940. But the time was right to mass-produce furniture for middle America, and Clyde Harris stepped up to the challenge. Soon after the church took over Harris Pine, unaccountably the Chicago's American Furniture Mart assigned the company a prominent double space to showcase their wares.

My father's report to the 1964 General Conference audience explained, in short, the church's presence at the furniture mart—an outgrowth of the

[1] "Furniture at Mart," *Time*, July 9, 1934, http://content.time.com/time/magazine/article/0,9171,769942,00.html.

expansion plans going on at Harris. With their output and sales in the years following, Harris Pine would challenge the major furniture makers.

During the growth years of Harris Pine Mills, the marketplace, the center of Chicago, indeed at the center of the nation, took on importance in the selling of furniture. At first a twice yearly event, it was followed by a year-round showroom exhibiting the newest and best of America's furniture.

And so it was that Harris Pine's corner space directly beyond the second floor elevator was affectionately known as "the church." There was camaraderie in the multistoried building. The Far West company, with its array of furniture displayed with dignity, if not with the usual design-team-from-New York flair, caught the attention of Jewish merchants as they stepped off the elevator.

As imperceptibly, akin to when Jesus walked in Jerusalem, the church took up residence in the halls of commercial enterprise. Amid the noisy business, as merchants negotiated their wares up the elevators and made deals with retailers, stood this conspicuous corner space that invited questions. *Who are they? Where is Pendleton, Oregon? Who is Harris? Oh yes, the couple in the news who gave their business to the church.*

The church was its identifying feature. Not that they weren't interested in the merchandise. Sales ramped up and put the home plant in the stretch. Adding branch plants in locations near Adventist schools would alleviate the crunch as parts were shipped out to be assembled and finished from multiple locations.

Yet the multipronged task of finding land, building, developing, and staffing branch plants, to say nothing of producing a marketable product, was the daunting task mill executives faced. The velocity at Harris Pine Mills spun way beyond even Clyde's wildest hopes and dreams. And the end product of the church's furniture company landed at the corner space in Chicago's furniture mart. There were other furniture markets around the country, but with its Eastern connections with the biggest merchants, Chicago would remain the most important.

The church had 4,000 feet of floor space to display their products. As the elevator doors opened on the second floor, the first floor of showrooms, Harris Pine Mills, Inc., spread out before the spilling crowd, which was eager to get on with business.

But Saturday, the busiest day of the week, the church's showroom was dark. Elevator doors opened. Customers peered into shadowy and vague outlines. *This company wants to sell furniture and it's shuttered today?* Some would say it detracted from the spirit of merchandising. Dealmakers, their spirits high, paused at the church's space, the clamor muted if only for a moment. The cigar-smoking bigwigs guffawing their way up the elevator stopped amid their jokes. Eyebrows raised, buyers looked at one another, pointed, and shrugged. And went on their way to find the brightly lit spaces willing to do business.

But at the end of their business day they again passed the darkened showroom by the elevators that took them down to the street. Questions arose. On other days of the week the space beckoned them inside to run their hands over the soft-gleaming ponderosa pine. Soon other woods were added—beautiful maple and alder, and the nationwide favorite, the redwood outdoor furniture.

Admiration and curiosity drew buyers in with a question: "Why are you closed on Saturday?"

The answer: "God gave the Sabbath as a gift to us. It enables us to rest, to remember our Maker, and give thanks."

Word got around. The cigars stopped puffing, if only for a moment. The women looked thoughtful. Someone said, "Ah yes, my mother taught me about the Sabbath in our Jewish home." They punched the elevator button and likely pondered things they hadn't thought about for years.

The Harris Pine space thrived. Business burgeoned as catalog companies embraced Harris of Pendleton. It wasn't all at once. Indeed, if one were not paying close attention you'd have missed the significance. But as Charlie came and went from Chicago, he noted more and more signs going up: "Closed Saturday." And when in the annals of the furniture mart, it closed on Saturday, the significance wasn't lost on Charlie.

It isn't possible, some would have said, for an unknown furniture company in a Western cowboy town to become first in furniture manufacturing for Middle America. But the merchants got it. The church, respected for its values, beamed a light into the hearts of the merchants.

The 666 address raised eyebrows, and not just among the pious. Years later, as the building changed from furniture mart to condominiums and Playboy Enterprises spilled across the entire top floor, discomfited saints and sinners quietly had the address changed to 680 Lakeshore Drive. And Harris Pine employees just smiled as Charlie had done when he unabashedly invited the 1964 General Conference attendees to visit the church's prophetic address in the middle of the halls of commerce.

God goes about getting noticed in the most fascinating places and ways.

SHENANIGANS AND BIG DEALS

CHAPTER 32

After signing over his company and taking a month's cruise viewing Alaska's icebergs and mountain ranges, Clyde was restless. He returned home declaring there was no prettier place on earth than the Wallowa valley, mountains, and lake. He had his eye on a 200-acre farm with a meandering brook through the wheat, just east of Joseph. With Mount Chief Joseph looming over the octagonal barn, he envisioned Mary's horses splendidly grazing the pastures and drinking from the stream that flowed through the property. For a mere $200,000 the couple could own the spread.

He would add to the real estate he already owned, lakefront he had purchased for a song in the 1930s when he'd built a logging road on a three-mile stretch along the wooded east side. Then he'd brought in a couple of logger shacks, built a catwalk between them, and called it a cabin. When the couple entertained friends, one side held Mary's wood stove, the water bucketed up from the lake. The other cabin was for communal sleeping in the curtained-off spaces. The outhouse was down the path, overlooking the lake.

Clyde would hire Mary's brother Harold Coe and his family to tend the 1,200-acre wheat ranch. And he made plans to build a real cottage on the lake. The curiosity for those who visited the new site was the thick cement bomb shelter plunged deep into the embankment behind the cabin. It was Clyde's solution to the nuclear threat so real in the 1950s. In a primitive, dank, cold 8' x 10' room, Clyde made provision to shelter Mary in case of an atomic attack.

Over time the bomb shelter became a storage place for rusty corkscrew bedsprings, single rowboat oars, and faded orange canvas life jackets. As the fears of nuclear threats waned Clyde cut a window into the 12-inch-thick concrete wall facing the lake.

Clyde fished so often and so long that the Wallowa County game wardens went on the alert. Who could miss the dusty Cadillac or his red

HPM pickup cruising those back roads by the lake and Big Sheep and Little Sheep creeks? The game wardens suspected he was pulling out more than the daily limit of fish and tried, unsuccessfully, to catch him red-handed. Clyde, tired of the harassment, treated them with a curtness he usually reserved for phone conversations. That didn't set well with the upholders of the law. He was suspect, and the lawman who could document it would be acclaimed a hero by the Department of Fish and Game in Enterprise, Oregon.

The day a young warden fell in behind Clyde's red pickup and followed him to the cabin's gravel parking lot, past the unlocked gate, was the day this game official would deeply regret. "Mr. Harris, I need to check your catch for the day," the warden said as he stepped from his truck.

Clyde shrugged and let the rookie search his fishing creel, as well as inside, under the seats, and the trunk bed. Finding only the legal limit of 10 trout, albeit big ones, in Clyde's creel, but suspecting more, the young warden announced he'd have a look inside the cabin.

Gathering together his fishing gear, rods, vest, and creel, Clyde shrugged. "Suit yourself."

The warden was methodical, paying close attention to what was in the refrigerator and the freezer. Meanwhile Clyde jumped into his little red truck, whirled in the gravel, and disappeared back down the dirt road. He would be as methodical as the snooping warden. A half mile to the north end of his private stretch of lake property, he swung the gate shut, wrapped the sturdy chain around the steel post, and padlocked it. Then he casually made his way to the dock at the lake's edge, untied his 35-horsepower outboard, pulled the cord, and, motor purring, steered the outboard to the middle of the mile-wide lake. There he cast a fishing line and began to troll.

The game warden reappeared empty-handed, having found no evidence. He hopped into his rig and headed out. Not this time, but soon, he reasoned, he'd catch Clyde Harris with the evidence.

The padlocked gate came as a shock. Just minutes ago it was unlocked. The game warden panicked. The rest of the afternoon there was a lot of activity at Clyde Harris's property, particularly around the padlocked gate. Radio contact to the Bureau of Land Management described an unusual predicament and why backup was needed with a pair of bolt cutters on private property ablaze with "No Trespassing" signs. Only Clyde's friends were let in on the story to appreciative chuckles. There was complete silence from the Department of Fish and Game. Clyde didn't hate the laws of the county; it was the harassment that irked him. And to the Bureau of Land Management's credit, they left Mr. Harris alone after that.

Well into the 1960s Clyde still envisioned big deals. He had his farm and his Wallowa Lake property. And he still held control of Milton Box Corporation, which held timber parcels that he sold over time to Harris Pine. But in 1960 he plunged into the beef cattle feeding business, designing a pellet of feed that promised to simplify the feeding of thousands of heads of cattle and sheep at a time.

The feedlot deal began in 1959 and lasted well into the 1960s. Compressed feed, if you will, would simplify feeding and fatten up the livestock for a big profit. It took a $1.1 million loan from A. L. Powers to develop the process. Clyde raised interest rates on the timber he still owned in Milton Box Company. It would also require additional HPM loans of cash. Then, to protect himself if his health should fail, he appointed Charlie Nagele his successor in the business.

The big deal strained the credulity and resources of First National Bank and Harris Pine Mills, to say nothing of the brethren who came along for the ride, so to speak, conjoined with Mr. Harris by their mutual association with Harris Pine Mills.

The business was heralded in eastern Oregon and Washington with full page welcoming ads and reports of Clyde's newest venture in eastern Oregon. Welcoming this as the future of the beef cattle business, Clyde envisioned his newest venture skyrocketing just as Harris Pine had done.

The details he chose to overlook were (a) church leadership was not exactly enthusiastic about the enterprise, (b) Charlie Nagele began taking aspirin, and (c) A. L. Powers, though a strong believer in Harris's business acumen, announced that the $1.1 million loan was as far as U.S. National Bank would go in the cattle-fattening business.

And there was one other thing, something Clyde ignored. The cattle didn't much like the feed trough pellets. What was wrong with hay? they mooed. But Clyde was too busy building his new industry to listen. And to hear Clyde tell it, the new company's time had come.

The newspapers, *Union Bulletin* of Walla Walla and *East Oregonian* of Pendleton, devoted three pages to a welcome and story about this businessman who was magically successful.

On March 30, 1960, the *East Oregonian* reported that Clyde Harris was "now operating a million-dollar push-button livestock feed processing plant at Barrett Station, near Milton-Freewater" and that he thus expected to see a big meat packing plant established in the area.

"If such a plant becomes a reality," said Clyde, as quoted in the article, "it will be possible to expand many phases of livestock production and processing in this region. Finished livestock would no longer have to be shipped out of the county for processing."[1]

The article further describes Harris's new business:

"Before he embarked on his present venture, Harris visited feed processing plants in Oregon, Washington, Colorado, and Utah and incorporated the best features of each into his Milton-Freewater project.

"At Milton-Freewater he has two pellet mills and one rolling mill which can be operated by two men on a shift. The plant turns out 120 tons of pellets, cubes, and bulk feed in 20 hours. According to Harris, his are the only livestock feedlot mills in the West where so many different mixing and batch sequences can be accomplished. . . .

"At the present time, Harris is finishing out 4,500 lambs and 1,000 beef cattle in the vast feedlot adjacent to his plant. He plans to soon be feeding at least 10,000 cattle and 10,000 lambs. Within 10 years he expects to furnish pellets, cubes, and bulk feed to approximately 10 Oregon and Washington major cattle and lamb feeders."

An adjoining article, "Harris Will Employ 30," included additional particulars about the operation:

"Harris explained that all feeding equipment for fattening lambs and cattle is mechanized. Cattle feeding mangers range from 1,000 to 2,000 feet in length. Lambs are fed pellets from self-feeders. Concrete aprons have been erected behind all cattle mangers," the article stated.

"During the summer months, Harris utilizes a 1,000-acre irrigated pasture near Joseph, Oregon, for the cattle-feeding operation. There yearling calves gain as much as two and a half pounds a day on pasture grass with grain supplement. The pasture is located high in the foothills of the Wallowa Mountains."[2]

Clyde needed loans and cash for his latest venture. It was natural that he would cast his eyes to his old company and moneymaker, Harris Pine Mills. He began figuring more closely on the 1951 agreement with the brethren. He upped the interest rate as he sold his timber holdings to

---

[1] "Earnings From Acre Give Harris Start," *East Oregonian*, Mar. 30, 1960, p. 2.

[2] "Harris Will Employ 30," *East Oregonian*, Mar. 30, 1960, p. 2.

Harris Pine. He changed the scheduled timber sales and threatened to sell elsewhere if HPM couldn't comply with his new schedule. When he was cash-poor, he pushed for more. If the officers at Harris Pine Mills argued with his changes, he inferred he would sell the timber to another company.

Sometimes when Charlie returned to his office after lunch he'd find Clyde sitting in the old leather chair, a check for $50,000 in one hand (the profit on wheat), the phone in his other hand, spending the money on fertilizer and improvements for his land. He was always short on funds.

Annually the General Conference paid a $25,000 annuity to Clyde and Mary, as agreed upon in 1951. This would continue until the deaths of both of them. Clyde kept Mary on a strict budget and put away large amounts of money for his business deals. Mary never got over feeling poor. Her checkbook reflected very little of their steady income.

Charlie recorded Clyde's plans in the form of notes. "Idea would be for CJN to oversee the whole operation eventually after CHH fully retires. CHH will write to General Conference making this suggestion," Charlie unhappily jotted down.

"The feeding plant, Barrett's station, put a strain on Harris Pine. The dialogue between Clyde, the banks, CPA Fred Niemi, and Charlie Nagele dragged on. But then came a dramatic confrontation in Portland, Oregon, with HPM attorneys, Fred Neimi, CPA, and longtime friend of the Harris couple, A. L. Powers, the U.S. National Bank president who had held faith with Clyde during the development years and covered the 1951 shortfall when the payroll checks had already been handed out."

Another of Charlie's notes (dated October 30, 1959) stated the following: "[CHH] also suggested CJN get acquainted with cattle operation finance and help find a manager for cattle and pellet operation so in case anything happens to CHH, General Conference under CJN could take over in harmony with will to operate with Mary. Would look to CJN to oversee."

On January 3, 1963, Charlie took out his notepad and recorded further relevant facts:

"CHH having some difficulty with U.S. National Bank on credit line for Milton Box Company and Harris Milling and Feeding. [CJN] went to Portland twice for him, urged bank to extend $1,100,000. They agreed—provided

- Milton Box Company and Harris Milling and Feeding Company were merged
- $50,000 note at Oregon Bank was paid off
- HPM advance an additional $100,000 on Blalock and Burlingame

timber to improve cost position of Harris Milling and Feeding Company to permit the purchase of more cattle and paying off of $50,000 note to at Oregon Bank

"Harris Pine Mills agreed to advance an additional $100,000 on timber contract on the following conditions and promises from CHH:

- all proceeds from timber cutting beyond 1963 be applied on advances
- understood all at $10 per thousand because of interest factor

"CHH agreed that all cutting would be done by HPM and that he would refrain from putting pressure on CJN in the future when he was in need of money (cash) (pressure by inferred threat to sell to others). He [Clyde] agreed and said he would not in the future and understood that no more advances would be forthcoming. He also agreed . . . that in all probability $1,100,000 credit line from U.S. National Bank was maximum.

"CJN told CHH that the above was an absolute must and pressure at intervals upsetting past agreements must cease as CJN's health would not stand continuation of worry in this regard."[3]

The true measure of the man came through. My father's final notation was "CHH indicated definitely an agreement and said, 'You have my integrity in this to bank on.'"

Clyde stood by his word. All his life he had been plunging and looking ahead. But now he recognized that he'd come to the end of his big deals. No more expanding the beef cattle business. Shortly thereafter he sold his wheat ranch. It was a sad time in his life, but he still reminisced about how the Lord had led him in the gift of Harris Pine Mills. He looked at the branch plants, toured them from time to time, and reveled in the success of student labor and the kids getting their education through Harris Pine Mills work. And he pronounced it good.

How hard it is to let go. Maybe the prophet Elijah knew a little about that as he sat in the cave after his 40-day sprint across the desert. Age and human frailty had won. God met him in the still small voice, not the big deals of wind and fire, and told him to anoint Elisha. Elijah was to slow down and regard his waning strength. It didn't mean that God loved Elijah any less for his humanness. God knew. Time to take your hands off the steering wheel, Elijah.

Clyde lived two more years. Quietly. The pellet mill never reached the

[3] Charles J. Nagele, notes, in the author's possession (for all above dated notes).

glorious heights he had imagined through the mid-1960s. It was still in operation when he died, and the General Conference was the recipient of a feed yard and 5,000 head of beef cattle. The brethren coughed at the mother lode they had just inherited, and Charlie quickly cast around for a buyer. It saved some embarrassing publicity for the Seventh-day Adventists, often known as a people who tend to follow a vegetarian lifestyle.

"THE SUN ALWAYS SHINES ON BLALOCK MOUNTAIN"

CHAPTER 33

She was a legendary Annie Oakley—Claude Harris's youngest daughter and the delight of her outdoor-loving Uncle Clyde. She joined the Harris clan as Claude and Maude's third child. Early on, the family gave up trying to fence her in. She could shoot, fish, hike, hunt, as well as bag truffles better than any outdoorsman in the valley.

From the age of 10 Marcene Harris Garriot remembers riding horseback with her dad and tending his thousands of sheep grazing on their own property that stretched to the Columbia River.

Before Milton Box Company, Blalock Mountain belonged to her maternal grandfather. Her family enjoyed their wood-walled tent on Blalock Mountain long before Clyde horse-traded to become owner. They would steer their horses straight up the winding eight miles to the outdoor wonderland that Marcene believed was the top of the world. Hers is pioneer ancestry on both sides. Her mother's family arrived in the valley at approximately the same time as the Fraziers. "The Demaris family was considered upper crust," she says with a wry smile, adding, "if there was such a thing in the valley." Then again, if you inherit a mountain as Maude Demaris did, it does put you ahead of the rank and file citizen of any locale.

Details are hazy, but she remembers that her dad, Claude, gave or sold the mountain to his little brother Clyde and Milton Box Company. Marcene isn't sure money was exchanged. It was the Harris family's way of doing business with one another. And baby brother Clyde was usually the recipient of their goodwill. They would often forget to make him pony up the money he owed, and Marcene says her daddy didn't mind a bit. Under Clyde's ownership the tent was replaced with a cabin, built with Milton Box Company lumber, so his brothers and pals could join him in the wilderness.

Still compressed into Marcene's small frame is an energized bundle of optimism, a natural drive to get outside and hike the eastern Oregon wilderness. Her pioneer ancestors would quickly recognize her as one of their own. Unafraid of the wilds and things that go hoot in the night, she's a family legend. She loves being alone, hiking the canyons and hills of Blalock, hunting elk, fishing the streams, and letting the wind whip through her curly brown—these days, turned almost white—hair. While she's traded horseback for a Land Rover, she still treads dangerously to places the travel brochures don't suggest. But the mountain close to home has always been her playground, and truffle hunting is her particular skill. She enjoys the pursuit of those mysterious fungi for which epicureans the world over pay astonishing sums per pound.

For many years Marcene continued to hunt and fish on Blalock, often with her Uncle Clyde. But the excursions grew further apart. Then just a week before Thanksgiving 1968, Uncle Clyde unexpectedly called Marcene, who lived in the Milton Valley, to join him in a jaunt to the top of Blalock. She hesitated, thinking he was being impulsive. Uncle Clyde wasn't as vigorous as he used to be. The valley was shrouded in fog, and she knew the snow-laden incline would be hazardous, even for a four-wheel drive. But Clyde insisted they make the trip in his ancient little red pickup. She knew it wasn't heavy enough for the snowdrifts they'd encounter, but she'd never been accused of being a sissy, and she couldn't say no to her favorite uncle.

He insisted on driving. The snow was deep. And just as she figured, they got stuck right past the gate at the bottom of the mountain. But Clyde wouldn't give up. So, laughing at their intemperate pursuit, they dug the little rig out of the snow. Then heaving rocks into the truck's bed to add weight, they continued on inching through the fresh snow. The truck slipped and slid the entire eight miles with their gunning it through the drifts, Uncle Clyde laughing like a kid.

Marcene remembers that he talked nonstop the whole time up Blalock and back. The view at the top was worth all the skidding. The sun-drenched snow sparkled like acres of Tiffany diamonds. The cabin and the forests beyond outdid Thomas Kincaid's best work. Clyde tromped happily through the snow. He unlocked the cabin, looked around, and declared that everything looked fine. Then, on the brow of the mountain, where he used to perch with his shotgun aimed for the buck Charlie would surely scare up from the ravine below, he gazed for a long time into the fog-shrouded valley below, enveloped in the sun's rays.

On the way home he stopped at a roadside stand in the valley and bought a 100-pound sack of Walla Walla winter onions, the last of the autumn crop—as a peace offering to Mary, who worried constantly about his unadvised activities. When she learned of his risky adventure and saw his contrite offering of bulging burlap, she shot a glance at the bag and said, "What in the world am I going to do with 100 pounds of onions?"

One evening my parents stopped by the Harris home. It was few days before the 1968 Thanksgiving holiday, and they were heading to Sacramento to join my family. Clyde wasn't feeling well and sat in his white leather chair, saying he had a cold. He seemed vulnerable and pensive. My mother, always the nurse, took his blood pressure. It was too high. When she leaned down to kiss him goodbye, her lips brushing his forehead, she thought he felt feverish. Maybe he had a touch of the flu, she thought. It was with a sense of unease that my father and mother said farewell to Clyde and Mary. They comforted themselves that Dr. Easton, who lived next door, and Dr. John DeRomanett, Margaret's husband, were always standing by.

My parents arrived in Sacramento late Wednesday evening. The sun had barely risen Thursday morning when the telephone roused us out of slumber. "Uncle Clyde died this morning" was Margaret's tearful announcement. No one was prepared for this. Dr. Easton was delivering a baby in the predawn hours. Dr. DeRomanett had taken a quick trip out of town to check out the frozen water pipes under his Wallowa Lake cabin.

The news struck my parents like a thunderbolt. Forgetting their suitcases, 10 minutes later they were in their car and backing out of our driveway. The *East Oregonian* notice seemed too sparse to those who honored and loved him: "Clyde Harris, 78, who gave a multimillion-dollar business to the Seventh-day Adventist Church, died in his home here Thursday morning. Friends of the family said Harris had complained of a minor discomfort and collapsed while his wife was getting an aspirin for him. Services are tentatively set for Monday at the Adventist church here, with Burns Funeral Home handling arrangements."

Clyde had awakened in some distress at 4:00 a.m. He thought a little potato soup might settle his stomach, so Mary went to the kitchen to heat some. As she stirred the kettle, her thoughts went to the bountiful Thanksgiving dinner she was preparing. She made note of what time she should put the turkey that Clyde so enjoyed in the oven. Then hearing his call of distress from the bedroom, she dashed to his side. He was gone.

Marcene has no regrets about their adventure to Blalock the week before. It is her treasured memory of the man she admired and loved. She adds, "With the cold fog that settles in the Walla Walla Valley between November and March, it sometimes seems as if the sun has packed up and left for good."

So why did Clyde ask her to join him in his little red pickup and make the difficult drive to the top of the mountain that day?

"Because," Marcene says, "the sun always shines on Blalock Mountain."

THE OTHER CHARLIE

CHAPTER 34

It's a long trip by train from Cedar Grove, Tennessee, to Farragut, Idaho—especially if you're standing up all the way. A packed troop train yielded no seat for Charlie Fry the summer of 1943 when he, his wife Lucille, and 2-year-old Wendell headed West.

Born into the middle of six children, Charlie grew up in a Seventh-day Adventist home. The family was cotton farmers. He married Lucille, he says, "when we were still just kids." Charlie's older brother Hollon had already discovered the plentiful orchards of the Yakima Valley in Washington State. His letters home suggested that jobs were plentiful, and when he urged his siblings to join him, Charlie and Lucille took a chance. Eventually, the others would follow. Picking fruit in the fertile, cooler Northwest was reason enough to leave home and the back-breaking job of picking cotton.

Farragut, the newly created 4,000-acre naval base, gave basic training to the Blue Jackets. It was located at the southern tip of Idaho's Lake Pend Oreille, and there young men who'd left home for the first time learned how to march, row, swim, and use firearms before they were sent to the Mediterranean Sea or the South Pacific. At last the conductor produced a disembarking step for Charlie to perch upon. He was very grateful, but when he sat down, the rough points of the steel surface bit into his backside. The insult to his posterior or his buckling legs—he couldn't decide which was worse.

When the conductor announced, "Next stop, Farragut," the troops began gathering their gear, slinging duffels over their shoulders. Joshing each other to cover their fears of what lay ahead, they strained to see the first wispy view of the naval base that would house and train almost 300,000 servicemen before the war's end. As they spilled onto the platform, Charlie sunk onto a wooden seat.

Six hours later the train whistled into Toppenish, Washington, and

Charlie roused to see Hollon waving a welcome. His brother drove them to the Granger labor camp in a borrowed car. They arrived during cherry picking season, which was soon followed by peaches, plums, and apples. They worked alongside the legal (and illegal) immigrants who followed the yearly crops. Home for the little family was a one-room cabin, modern amenities lacking.

Granger, a tiny town in a fertile farming valley in central Washington, was a good place to settle down, but Uncle Sam had other plans. They'd gone back to Tennessee to await Charlie's draft assignment, when suddenly the war was over.

So they went back to Granger and were about to endure another jobless winter, when something important happened. Charlie saw one of the ads Clyde Harris placed in the back of the *North Pacific Union Gleaner.* It was just one line: "Workers needed. Apply at Harris Pine Mills, Pendleton, Oregon."

So it was that on a brisk October morning in 1947 Charlie pushed open the door at the head office of Harris Pine Mills. Quick-moving, perceptive Bud Smith, Mr. Harris's nephew, controlled the hiring at Harris Pine Mills, and took a liking to Charlie. With a distinct lack of formality or orientation, Bud rustled paperwork to put him right to work on Assembly I. Charlie fidgeted, wondering how he was going to get his family to town. He stole a glance at the far end of the knotty pine door with its stenciled frosted window: Clyde Harris, president.

What must it be like to sit in the president's chair? he wondered.

"Your starting pay is 60 cents an hour," Bud said, the paperwork over.

So Charlie went to work. He didn't know where he was going to sleep that night, but as he walked out the gate at quitting time, someone mentioned that eight guys were batching in a big house a few blocks away on Byers Street. Charlie inquired further. The jovial fellows invited him to be the ninth housemate. They had almost no cooking and cleaning skills among them, and every morning all nine lined up for the one bathroom. Even now, five decades later, Charlie smiles. "You get the picture," he says.

Lucille and Wendell came to town to rescue Daddy from the Byers Street bachelor pad, but there were no houses to rent. The war, on the heels of the Great Depression, had stalled building and depleted rentals. As Harris Pine Mills expanded, the situation was bleak. Clyde and Mary Harris were knee-deep building 80-plus bungalows on South Hill, hoping to alleviate the crunch and offer affordable housing to families moving to town. But that was above the Frys' budget.

"Lucille was a trooper," Charlie remembers of his wife's dogged pursuit to keep the family together. "Every day she walked up and down the streets asking for something to rent. When she finally found a lead, it was a back bedroom. "One-room living is cramped," Charlie says dryly.

Loading lumber into boxcars wasn't Charlie Fry's idea of a dream job promotion, but Bud Smith thought it was. Charlie was moved outside. This young go-getter would do well loading lumber into the boxcars on the railroad siding. As Charlie sweated the heavy lifting, Marvin Robinson, the straddle buggy driver, deposited load after load of lumber without breaking a sweat.

"Those hyster and straddle buggy drivers have it made," Charlie said to himself. Moving loads of lumber around the yard with a straddle buggy or hyster—now, that was a job worth aiming for. "I thought, *If I could get Marvin Robinson's job, I'd be happy the rest of my life,*" Charlie said.

Gossip floated around that Marvin might be promoted to foreman of Assembly I, so Charlie sidled up to Marvin at break time. "How about putting in a word to Bud for me as your replacement?" he said.

"I'll do it," Marvin said.

Nothing happened except that Marvin was made foreman. The morning that Charlie saw Warren Freund climb into Marvin's straddle buggy was a bad day for Charlie.

"I have something else in mind for you," Bud said one day in his usual clipped way. Like his uncle, he didn't do small talk, ever. The family resemblance was strong. And yes, Bud knew that Charlie aspired to Marvin's straddle buggy job.

Bud appeared a couple of days later and motioned for Charlie to follow him into the box factory. They stopped at Assembly II, and Bud indicated that this was Charlie's promotion. *This is a promotion?* Charlie thought as his spirits dropped into his work boots. Warren had the only job he aspired to, and here he was back on an assembly line. But he bent into the new job, and months rolled by.

It was just before Christmas in 1950 that Bud strolled by Charlie at Assembly II. With no preamble, Bud said, "We are thinking of making you the manager of the Alexandria branch plant. You'll have to move to

Virginia." Harris Pine's first branch plant in the renovated brewery was Gib Garver's baby, but Clyde had his eye on Garver to build a Chicago Heights plant to act as a central United States assembly point.

Charlie stood speechless as Bud disappeared around a machine. "That was all," Charlie remembers. "I'd been inside the head office once, and that was the day Bud hired me."

We are thinking—the phrase Bud used with this new proposition—rolled around in Charlie's head. Who were the "we" of the company who were giving him this promotion? He had never met Mr. Harris. Fry was used to seeing Clyde stroll briskly through the mill, but Mr. Harris never paused. Charlie pondered how "we" were promoting him.

Another two weeks rolled around before Bud reappeared. "Things are moving fast. Better get ready. We'd like you to go middle of this week."

"This *week*? What about my family and furniture?"

"Well," Bud said, "we'll give you a couple of days. Just load your household goods into a boxcar with a load of shook heading east."

And Bud disappeared around the corner.

So 29-year-old Charlie Fry, with no experience and no introduction to the company president, now managed the Alexandria branch of Harris Pine Mills. Twenty employees looked to him for leadership.

Charlie might have panicked had it not been for Ethel, the capable secretary from Harris Pine's Pendleton office, sent to the fledgling startup branch in the East with the Garvers. Ethel Taylor was Garver's secretary, and she would teach Fry how to run a business.

"She saved my hide," Charlie reflects with amusement. "She taught me how to dictate letters, act like a boss whether I felt like it or not, and how to fire an employee when the need arose." But Miss Taylor's tutelage was short-lived. She soon announced her engagement and upcoming wedding. After all, these were the 1950s, and attractive, competent women who'd grown up playing the card game Old Maid knew that a marriage proposal always trumped a career.

By the summer of 1951 Charlie Fry was getting his feet on the ground after the quantum jump from assembly employee to boss. So the morning that two men wearing suits and ties showed up at his business and announced they had come to take a plant inventory for the Adventist General Conference, Charlie said, "Wait just a minute."

He offered them a seat just inside the front door while he placed a call to the head office in Pendleton. The brethren who were asked to cool their heels happened to be Charles Nagele, treasurer of the North Pacific Union, and Arthur Spoo, business manager of Walla Walla College. Although the Harrises' gift to the church was only in the discussion stage, the two men had been commissioned by the General Conference and Harris Pine Mills to travel to Alexandria and inventory the plant. No one had thought to notify Fry.

So Fry picked up the phone.

"Number, please," the telephone operator crooned.

"Operator, I'd like to place a person-to-person call to Pendleton, Oregon, to Mr. Bud Smith at Harris Pine Mills." He then gave her the number.

"One moment, please."

Fry waited.

Bud came on the line.

"Mr. Smith, this is Charlie Fry. I have a strange situation this morning. Two men are sitting here saying that the General Conference sent them to inventory my plant. Do I let them in?"

"Yup," Bud responded. "Clyde's working on a deal with the denomination, but I don't have any more details than that. Let 'em have the run of the place and give 'em what they ask for." That was all.

So Mr. Harris was thinking of giving the mill to the church? Fry was getting used to surprises at Harris Pine.

For seven and a half years the Fry family lived at Alexandria. Charlie Fry became acquainted with Charlie Nagele and the church officers at the General Conference headquarters in Silver Spring, Maryland. He even hired one of the General Conference secretaries, Miss Bacon, to replace Miss Taylor. Periodically he was ribbed about "frying Bacon" over at the Alexandria HPM plant.

Ten years after leaving his job on Assembly II, Charlie was building and managing the HPM branch near the new Blue Mountain Academy in Pennsylvania. One morning in 1961 Charlie Nagele and Charlie Fry slid into a booth at a diner in Hamburg, Pennsylvania. Over their hash browns and eggs Nagele told Fry, "I want you to consider moving back to Pendleton as vice president in charge of building and developing the new plants as they come into the company. We need a level head, someone with good business sense to integrate the new branches into HPM's product lines."

The company was adding plants at a breakneck speed. Many Adventist schools were in financial straits. While denominational schools provided on-campus jobs so their students could earn part of their tuition fees, they were paid out of funds that came from the institutions themselves. The schools needed outside industries. Many schools had already attempted production plants of various products with some success. But from wood products to upholstered furniture, one by one many Adventist academies and colleges turned to HPM for help.

Many in conference leadership lobbied hard for Harris Pine to become their miracle worker: "We're in trouble. Please take over our industry." All over the United States the plea was "Can you build us a plant?"

The schools begged, Union and the Conference officers cajoled, and the schools scooted aside to provide land for HPM plants to be built nearby.

Harvard Business School didn't have a course for absorbing nonprofit school enterprises into a for-profit corporation at the rate and speed necessary to help educational institutions. It defied logic and good reason. But Charlie Nagele purposed to create jobs for Adventist education. Every year a handsome figure out of the company's abundant profit was turned into the church's mission endeavors. The bottom line proved that this was more than a human endeavor.

As Nagele talked and Fry poked at his scrambled eggs, again astonished, he couldn't help thinking back to the day Warren Freund climbed into the straddle buggy and Fry watched his highest ambition snatched away. He marveled at the unexpected turns his life had taken. Now not only would he get to see the inside of the head office, but one day his office would be down the hall and behind the frosted window he'd glimpsed that October day when Bud Smith hired him for Assembly I.

When Nagele retired, someone asked him if he'd ever envisioned being made president of the company? "Not in my wildest dreams! I'd have said you're crazy!"

But this wise man became the leader of a huge industry, on first name terms with Montgomery Ward's irascible old Sewell Avery. He rubbed shoulders with General Conference presidents, enjoyed a cordial friendship with Clyde and Mary Harris, served on operating committees, and became every inch an able leader. Humble roots can be the best way to prepare for great things. You're teachable. And God loves that.

WALT A.

CHAPTER 35

He didn't have the brawn of Wild Bill, but he carried himself with a quiet dignity that commanded respect around the sawmill. When orders came from the front office to handle things a certain way, given his years of experience, Walt might point out problems with the changes, but he never went against orders. When time proved his point for him, he wasn't one to say I told you so.

As I grew up, I knew Mr. Applegate as a shy, soft-spoken head deacon who took responsibility at all church events, including Saturday night skating in the gym. At church he rarely sat down. He was always on his feet making room for visitors, collecting the offering, ever alert to making things run smoothly. But for more than 20 years he held the position of sawmill superintendent. He had the respect of the other bosses at the mill. Everyone genuinely liked him. Again and again my father remarked, "Walt Applegate is the salt of the earth."

He grew up on a cattle ranch near Weiser, Idaho. His father was known for his business acumen and honesty. It wasn't until his later years that Walt's father lost his fortune to drink. Walt left home when he was 12 years old—it was not unusual in that culture for a boy to leave home at an early age. Like Bill Tapley, Walt's first job was herding sheep. But even a pioneer's kid is still a boy at age 12, and Walt would admit that more than once he cried himself to sleep under the stars.

He learned mechanics by repairing trains in a roundhouse. When provoked, his impulse was to put up his dukes. Evenings, he learned how to box. In fact, his quick footwork wore through the soles of his bare feet, and an ensuing infection almost took his life.

In his early 20s he was hired on at Bates Mill, an out-of–the-way logging camp in the middle of the Oregon timber wilderness. He had briefly courted a pretty dark-haired girl in Weiser, but with an independent spirit bred into him by his pioneer ancestors, Walt left Idaho without making up his mind about marriage. One day Lois surprised him by stepping off a bus in Bates, where the infrastructure—one logging camp, one rooming house, and a post office—was indicative of the population. She didn't have a return ticket.

Walt secured a room for her in the boarding house, and Lois worked as a waitress in the small restaurant downstairs. Their romance quickly became an engagement, then a wedding. They were married in the small dining room, chairs pushed out of the way so the loggers could clump in and witness the nuptials.

Their first home was a one-room cabin in the middle of the logging camp. From there, that first winter, they slogged uphill through snow to the settlement. In the spring there were deep ruts and mud to deal with. Word spread that the village wino, wandering on a trail back to the clearing, was downed and killed by a cougar. Hearing that, Lois shivered in fear as she hiked the path to her waitressing job.

Lois had been only 13 when her mother died after a long, lingering illness. She had grown up in the Seventh-day Adventist Church but for a time did not live what she knew. Yet after she married, Lois shared the spirituality of her childhood with Walt, and the couple were baptized. They moved to Caldwell, Idaho, and for a few years Walter drove a bread route for the Gem State Academy bakery. Then he found a job doing what he loved, working on a ranch. His son Rodney remembers, "My father got the job of ranch hand on a 1,000-acre spread near Arlington, Oregon, but my folks wanted my sister and me to go to church school. Mr. Harris of Harris Pine offered Dad a job, so we moved to Pendleton."

The family arrived there the summer of 1942, and Clyde Harris put Walt to work repairing the sawmill trucks and hysters. Rodney reflects, "The old machinery shop was tucked in beside the log pond near the sawmill. I remember Dad coming home after 12-hour days working on the trucks and hysters. You couldn't buy parts in those days, during the war, so Dad made parts out of whatever scraps of metal he could get his hands on. I used to go down after school and watch as he squatted over a gallon bucket of carbon tetrachloride washing the grease out of the wheel bearings and other parts with his bare hands."

Shy he might have seemed, but Walt was a storytelling legend among the old-timers. During the 1940s many Saturday nights the church members gathered at the Applegates. Some brought root beer. They made homemade ice cream, the men trading off turning the old-fashioned hand crank. The kids played hide-and-seek under the streetlight, scattering to hide from Teddy, the border collie who was always "it." The kids hid while Teddy chased a ball or stick; then the clever dog systematically routed out every last one of them. If anyone cheated by hiding again before the ball was rethrown, Teddy nipped his or her heels all the way back to the streetlight. While they ate ice cream and popcorn and apples, Walt Bain and Walt Applegate regaled the folks with tall tales of growing up in the Wild West and Minnesota.

There was the tale about the thirsty hitchhiking prohibition agent Walt picked up one day on a lonely Idaho gravel road. He quenched his thirst on Walt's libation, then pulled out his badge. Walt left the guy on the lonesome gravel road without his shoes, miles from nowhere to walk home.

But Walt's favorite story was about Clyde Harris.

"It was a cold January day a couple years after I came to Harris Pine Mills," he'd recall. "I was trying to make a part and couldn't figure out how to do it, so I got out a repair manual and was boning up on how to repair an old truck transmission. I had my feet propped up on a 100-gallon can of crankcase oil with a big parts-and-instruction book in my lap. It was around 3:00 in the afternoon.

"Well, Clyde Harris strode in just then and stopped. I turned red, slammed the book shut with a bang, and jumped to my feet. I really thought he was going to fire me for sitting down on the job. So I covered my embarrassment with a joke. 'Mr. Harris, how 'bout a raise?' "

There was one thing all employees soon learned about Clyde Harris: he didn't miss much as he strode through his mill.

"I guess the place looked well taken care of," Walt would continue. "Clyde gave me a half smile and said, 'OK, go to the office and tell Arnold [the mill treasurer] I sent you for a raise.' "

He always hooted as he served the punch line.

His son Rodney adds an additional story that reflects the boss–worker camaraderie that sprang up between Clyde and his dad.

"One Christmas my dad came home from work with a gift Mr. Harris had given him. Dad was incredulous. It was a special set of micrometer tools, impossible to buy during the war and very expensive. It made the job of repairing engines infinitely easier, and the tools were his to keep. It was an unforgettable event in the Applegate house."

Then Harris Pine tapped Walt to be superintendent of the sawmill, a position he would hold until his retirement, more than 20 years later.

It wasn't his fault that a broken back landed him in the hospital. He couldn't have seen it coming. He was feeding boards into the edger when a malfunction occurred in the steam-powered lever. It kicked the board into the air and landed it on top of two high-speed blades. Walt was rammed against a wall 20 feet away. He was in the hospital a long time.

When Nagele sent Walt a hard-of-hearing young man with a hard luck story, Walt kept his eye on the guy even as he put him to work. It was a sympathetic gesture for a destitute family. With so much noise in the sawmill, the reasoning went, it didn't matter if one was hard of hearing. Those with perfectly good hearing couldn't hear out there either. But put a rookie hearing-impaired man in a new environment, throw in the stress of reading lips and hand signals, and Walt rightly sensed the danger inherent in a sawmill. The newcomer was an accident waiting to happen.

Walt put him on odd jobs of cleaning up trash and the ever-present sawdust that accumulates around a mill. The peace was soon interrupted by loud thrashing and yelling from the log pond. "The guy can't swim!" someone yelled, and jumped in to save the flailing man, who was hauled out choking and spitting the acrid water.

His next assignment was hauling chips from the log debarker to the boilers. There was nothing complicated about the safety of driving and dumping chips a block from the sawmill. But one morning Charlie Nagele was taking a walk around the mill at the time the deaf worker was hauling a load of trash. After dumping the load, he forgot to lower the bed. The rig caught on an overhanging catwalk and large blower pipe that took sawdust to the boilers, and cantilevered the cab high in midair. There he hung, looking down at Mr. Nagele, befuddled by what had just happened and what to do next.

Walt called factory superintendent Cloyd Wysong that morning. "I've got a guy over here that needs a job. Put him somewhere real safe."

In 1959 Walt became my gracious father-in-law. Even though he was more of an arm's-length, shake–hands, rather than an arms-around-you kind of guy, he was unfailingly pleasant. He didn't offer his opinions unless asked, and he loved the Lord with his whole heart. Get him started on his upbringing and his gentle mother, his wayward youth, and meeting Lois, falling in love, and embracing Christianity, and he would come alive. Like the garden he grew wherever they lived, the things he put his hands to always flourished.

Every year Walter planted a huge garden on Wild Horse Road, where they lived in their later years. The family ate in the best Adventist tradition—fruits, nuts, and grains with a heaping share of vegetables from Walter's garden. Lois could concoct vegetarian versions of almost anything. She made her own mayonnaise long before vegan cookbooks showed how to do it. She pioneered the recipes in her kitchen. No one ate at the Applegate table and came away hungry. The corn on the cob platter was piled high with Walter's best jubilee corn. Lois's salads, fresh from the garden, were legendary. And there were sprouts and whole wheat bread made from freshly ground wheat, still warm from the oven.

Walter was proud of Lois's cooking. She knew the health message inside and out and followed it closely. He enjoyed fresh carrot juice as much as she did. He agreed that meals should be five hours apart. He never drank liquids during his meals. In short, he was a true health reformer.

After his retirement from the heavy lifting of the sawmill position, Walt took on part-time mill duties, odd jobs, and work as a night watchman. He always took his lunch break at the same time each day. Weather permitting, he'd find a lumber pile to sit on while he ate his nourishing brown bag lunch. And he never let on that there was one tiny little thing that came between him and perfection.

That would be butter horns—those meltingly delicious pastries found in the grocery deli, probably not the first food that comes to mind when talking healthful eating. At lunchtime he'd eat his fat homemade sandwiches—Worthington Choplets, sprouts, and Lois's own eggless mayonnaise between bread made from freshly ground Montana hard wheat. Then after finishing with a crisp Red Delicious apple, Walter would stroll by the head office and out the gate. Lilly Hathaway always took notice. She had great affection for this soft-spoken man. He always returned with an Albertson's delicatessen bag in his hand and pure delight on his face. While he perched on the lumber pile and munched on butter horns, Lilly would smile to herself. His secret was safe with her. She'd never tell.

Walter lived to 95 and never lost his optimism. If one were to peruse his worn Bible, it would fall open to his favorite text, Psalm 27:14: "Wait on the Lord: be of good courage, and he shall strengthen thine heart: wait, I say, on the Lord." He read the text and through the years encouraged us many times. He was, to me, "the salt of the earth."

THE 15-MINUTE WALLA WALLA COLLEGE STUDENT

CHAPTER 36

Ray Watson wasn't considering a career at Harris Pine Mills when he showed up at the front office in 1951. He'd just transferred from a premed major at Union College. "I heard that students from Walla Walla College had a better chance of being accepted at Loma Linda, so my fiancé Pat and I traveled west." They arrived at Pat's sister's home in College Place, and by prearrangement with the college's maintenance supervisor, Ray went right to work. When the college business manager discovered I was a village student, he said, 'You can't work on campus if you live in the village.' So I was a 15-minute student at Walla Walla College.

"I went to Milton and worked in the fields for the summer," Ray continues. "But I couldn't stand the winter cold. Then a friend, Bruce Johnson, told me about Harris Pine MIlls but added the company wasn't hiring just then. I went to Pendleton anyway, found a seat in the waiting room and sat there. For three days. Then Bud Smith walked by, noting that I was still sitting there. "Can I help you?" he finally asked.

" 'I decided I'd be first in line when you begin to hire again,' I told him.

"Bud smiled. Probably only one of two times I ever saw him smile.

" 'When can you go to work?'

" 'Now.'

"Bud escorted me over to the dry sorter shed. And that was how . . . in 1951, I had the distinct privilege of unloading boxcars when the men in black overcoats [the Adventist officials] arrived," Ray says, enjoying the disparity of his heavy lifting job in the winter weather and the brethren shivering is they passed by on a plant tour escorted by Clyde Harris.

The couple were married at the Pendleton parsonage overlooking the wooden stairs leading to the Quonset hut gymnasium. "I worked at Harris Pine until 1952, when I went back to Union College and five more years of school."

In a providential coincidence, young Ray Watson and Reggie DeVice crossed paths. The Union College Furniture plant manager was the man responsible for designing the furniture for Clyde Harris and putting Harris Line of Super Pine on the map. When DeVice learned that Watson had just come from Harris Pine Mills, he said, "Have I got a job for you!"

Watson remembers, "It wasn't long until he offered me a job as a supervisor in the assembly department. I worked for him during the two years I was in school, and when he heard that I was out of funds to continue my schooling, he asked if I'd be interested in a full-time denominational job on the factory staff. This gave me security at a time my finances were at the bottom. It also included an opportunity to take some college classes each year. I appreciated working for him, and it gave me a chance to broaden my experience working with students. Mr. DeVice was complimentary and on several occasions encouraged me to consider this as a lifelong career."

It took young Ray a round-trip detour out west and work at Harris Pine to alter his plans for his life work. Under DeVice's influence, Ray changed his major from medicine to business. When the Oregon Conference asked Ray to manage the Laurelwood plant, the Watsons felt it was a call from the Lord.

"In a year I was on my way to the Monterey Bay Academy redwood plant to work with seasoned veterans Odien Shipler and Howard Marsh. When the brethren wanted to sell it to Harris Pine Mills, I was against it. It wasn't 'the work' to me. I was denominationally employed. I wanted to leave."

Ray says, "HPM bought it in 1960, and Charlie Nagele asked me to stay and run the plant, but I told him I wanted to work with students. Shepler went to the home plant in Pendleton. Charlie came to Monterey and said, 'Ray, give me two years. And then if you want to leave, I'll accept it. But with Harris Pine, you will still be doing the same thing, working with students and continuing to develop a product line, and you'll put a plant in Watsonville, as well as manage the one at the academy.' I was 29 years old when Charlie talked me into this. He convinced me that I had a future with Harris. He shared with me his vision for the company. Harris Pine began to open redwood and other branches all over the nation, and I saw the potential to grow."

One day Ray Watson would become vice president of Harris Pine. Seasoned in the wood products industry, he would guide Harris Pine to the top spot in America's outdoor furniture and number one supplier of outdoor furniture to retailers.

Little did Ray realize the contribution he would make to the denomination by accepting Charlie's challenge to come aboard Harris Pine. His years of service to the company were equaled by his meritorious service during the bankruptcy proceedings.

KANSAS GIRL

CHAPTER 37

She came from Independence, Kansas, with big aspirations for higher learning, but school tuition stood in her way. The summer that she worked at Enterprise Academy netted her $100, not enough to ensure she'd be able to graduate, so Lilly Miner traveled to Oregon in May of 1943. She landed a job at Harris Pine and a couch to sleep on in her brother's basement apartment. At summer's end she had a lot more than $100 in her pocket, enough to ensure her senior year, and a promise that she could return to Harris Pine in future summers.

But wishes, dreams, and aspirations take work to achieve. The HPM slicing machine, where water-softened wood was sliced into box shook, stacked on dollies, and dried in the kilns or the sun, had to be mastered. She took to the job with the careful skill of a perfectionist. Three other girls worked alongside her, sorting, slicing, and stacking, and after work they palled around together.

The next summer Lilly came West again, enrolling in Walla Walla College that fall as a business major. Trudging to the administration building's fourth floor and the business department was another challenge. But this clever Kansas girl was a trooper, and accounting came naturally to her. She earned her 140 words per minute shorthand and 70 words per minute typing pins, and was a whiz on the business machines. She worked as secretary to the department head, Mr. Cubley, and the registrar, Irene Black.

By the end of her sophomore year it was clear to her that she needed to work full-time all next year to keep abreast of tuition bills. Her parents had left Independence to come to Oregon and work at HPM, so Lilly moved in with her parents. Through the winter, oblivious to the plans her college teachers had for her, Lilly walked to work every day through the gate of Harris Pine Mill instead of up the four flights of stairs in the halls of higher learning. That is, except in bad weather, when Al Hathaway picked her up. Al and his wife were best friends with her brother Karl and his wife.

The day a car with Idaho license plates slid up beside her and a blond, blue-eyed passenger rolled down his window and asked her if she wanted a ride, she demurred.

"I've got just a little ways to go. I'll walk," she said.

At that she heard Al's familiar voice boom from the driver's side. "Why aren't you riding with me today, Lilly?" he asked. The blue-eyed stranger turned out to be Al's brother Don, just back from the service and on his way to apply for a job at the mill.

As Lilly slid into the back seat and Al introduced her to Don Hathaway, she thought, *H'mmm, he's cute*. Their first date was the Friday evening MV (Missionary Volunteer) meeting. They spent that Sabbath with other HPM couples at Emigrant Springs State Park in the Blue Mountains.

Lilly might have been tempted to believe she had passed up her calling when she learned about the plans Walla Walla College had for her. Anxious calls from Professor Cubley and Irene Black informed her they were awaiting her return. They had their eyes on her to become the next college registrar when Mrs. Black retired.

But she was not to be dissuaded. She married her wonderful guy in May 1947. The wedding was in Ed Paulson's backyard. Ed had been bragging that his lush new backyard needed a wedding—and bingo!—Lilly stepped up first. "Eight of us couples who worked at the mill were getting married that summer," she remembers.

It happened again and again at Harris Pine. Love interrupted the most carefully laid plans. The pretty Kansas girl with a watch on her right arm knew her ambitions had expanded and changed. And in God's grand scheme Lilly would reap the rewards of marriage, motherhood, and a career that surpassed all her ambitions. But that would unfold later. Now as the old "South Pacific" musical sang, she had fallen in love.

I'm as sunny as Kansas in August,
High as a flag on the Fourth of July!
If you'll excuse an expression I use,
I'm in love, I'm in love,
I'm in love with a wonderful guy!

During their first few years of marriage Don and Lilly moved about

11 times as Don tried various logging jobs around the Northwest. By 1956 they came back to Pendleton and Harris Pine for good. They had two little boys, and Lilly, a stay-at-home mom, felt her office skills getting rusty.

She decided to go back to work just long enough to get settled financially. She wanted to work in the box factory—where she wouldn't need her rusty secretarial skills. "Yet the office gave me an old-fashioned skills test, typing, numbers, and business acumen. On the test day I woke up with a fever, flu, and head cold and I did terrible. I failed the typing test. Although I did well on the machines, I was so embarrassed at failing the typing test that I avoided Mr. Johansen and didn't go into his office to meet him."

A few days later, she remembers, Joe Johansen, the mill treasurer, "one of the bigwigs," strode into the box factory. It was hard to miss the imposing, six-foot-four VIP. "He started over in my direction. I moved around the stack of boards. He kept coming around, and I kept avoiding him." Then he changed direction, and to Lilly's chagrin she ran smack into him.

Ignoring her effort to evade him, he said, "Lilly, I need you for the posting machine."

Her work in almost every department of the box factory had just ended. On the posting machine Lilly worked off five ledgers, blending them into one, and posting to payroll. She became the travel agent for the executives, the tedious job of searching the cheapest flights and rental cars into some very remote branch plants. When Charlie Fry assumed the job as company president, he appointed her as his secretary. Soon she was promoted to assistant sales manager and became expert at putting together the twice-yearly color catalog, assembling the tedious price list for innumerable items.

She made the job look easy, although it wasn't. "If I entered a wrong price, I sure heard about it," she says with a smile. "Recently I had a dream that the color brochures and price lists were stacked up on my desk and ready to send, when Charlie Fry came in and said, 'Lilly, there's been a price increase . . .'"

For more than 20 years Lilly was the editor of the monthly *Harris Line,* the employee news pages. She gathered interesting details from workers, and important dates and facts into a newsletter. Its front page, with a message from the president of the company, kept everyone informed and entertained.

Instead of blossoming in the hallowed halls of higher learning, Lilly's career blossomed in the midst of the rollicking, highly competitive tree-to-finished-furniture business that employed students who, just as she did, came to Harris Pine year after year to pay for their education. The Lord called her and blessed her with a host of talents that she used in His work. Lilly's skills made Harris Pine look its best.

THE ELEPHANT IN THE LIVING ROOM

CHAPTER 38

Each year from 1913, when Clyde and Dette started their small box factory, the business had prospered, with one exception. In 1933, in the midst of the Great Depression, the two founding brothers wrote off $75,000 in bad debts. That resulted in a company loss of $25,000. Clyde would always regret that decision, as it spoiled a perfect record of growth and handsome profits, including the years of church ownership.

After my father's retirement two successive executive teams held the reins at Harris Pine. In mid-1985 it was the General Conference president who replaced former General Conference treasurer Kenneth Emmerson as chair of the HPM board.

The early eighties begat a downturn in the economy as a general financial malaise hovered over the nation. Businesses were feeling the pinch as people tightened their pocketbooks. Aware that Harris Pine Mills must face the realities of the recession and the extremely high interest rates, mill executives began studying changes. The mill hired a consultant to advise them on how and where to cut. But their suggestions, many of which the company had already begun to initiate, didn't go to the heart of the problem. HPM needed to cut product lines, divest themselves of a few unprofitable plants, and suspend the addition of branch plants until the economy recovered. And church leadership needed to understand the economic dilemma and agree to cut back on branch plants for the sake of the company as a whole.

In the head office Lilly Hathaway noticed a tension in the chief operating officers. Yet her focus was pulled between her job and the office politics and a very personal struggle. Her husband had Parkinson's disease. She and Don had been married 37 years, and now the illness forced the love of her life to retire, and she was needed at home. In May 1986 she came to the decision that she would design the spring 1987 color catalog and verify that the price list was accurate when the furniture went to market in the new year. Then in March 1987 she would say farewell. Surrounded with new young faces in the office and the mounting stress, she felt convinced she had made the right decision.

In addition, the lumber and furniture industry was taking a beating over ecological issues. The spotted owl controversy and the downturn in the popular redwood patio furniture was leading customers to favor lighter-weight materials. It was difficult to stay relevant with the market changing so swiftly. Indeed, the entire wood industry was in turmoil, the nation in a recession, the economy forcing a depression in furniture sales, and Eastern buyers bidding timber sky-high.

"To confound their problems . . . in early 1986 . . . another local conference [Michigan] requested that HPM take over their two floundering plants. Harris Pine indicated they could not absorb the expenses of the two additional factories."[1]

In May 1986 the top two Harris Pine Mills executives approached the HPM board chair with a plan to cut overhead and reduce the number of plants. With their order file full, there was no need to cut production in the economic slump. But the proposal to downsize the company that for decades had been the lifeline of schools across the country didn't go over well.

The HPM executives were met with a swift pep talk: move out, expand, build inventory. Add more branch plants.

"Despite the fact that HPM had already pinpointed other financial and cash flow problem areas . . . , it was agreed by Harris management that given a couple of seasons, Michigan's business could feasibly be turned around. It was agreed that to include these two plants would be consistent with the mission assigned to them by the church."[2]

Harris Pine vice president of furniture, Clyde Iverson, recalls his own words to the HPM chair: "If we adopt the plan to move out instead of consolidating and closing unprofitable plants and cutting down the varied furniture designs, we have only a fifty-fifty chance of succeeding."

The Pendleton executives needed an infusion of cash to bridge them over the hard times if they were to take on two more strapped conference-owned academy businesses. Plus, changing styles of furniture demanded they create new designs to sell so the mill could remain relevant.

---

[1] Sandra Wachter-Van, "Harris Pine Mills: Some Personal Reflections," *Pacific Union Recorder*, May 4, 1987, p. 12.

[2] *Ibid.*

But when HPM president Spady and vice president Iverson explained how risky the expansion mode was in such a quirky market, they were reassured the church was behind them. "I can get you a loan of $35 million within two weeks. Let me go to work on it," the chair replied.

The two HPM executives left the meeting with assurances from the General Conference president that the church was behind them. Their orders were to move ahead.

At the home office in Pendleton, vice president Ray Watson, who helped mastermind the Monterey Bay Redwood Division, was delaying his usual midsummer redwood lumber buy for patio furniture scheduled to be in the stores by February 1987, until he received word that the General Conference was supportive of the "white water" Harris Pine was experiencing.

Ray Watson knew lumber and the redwood market inside and out. "It was a seasonal division," he remembers, "and during the production months we were producing 8,000 round picnic tables a day. I was holding up until I got the word to go ahead and plan for the next year's summer season. The brethren were going to make a loan."

Watson's job was to speculate and order enough redwood for the spring push into summer, then produce the product and get it into the stores. For six years the redwood division's popular patio furniture had kept Harris Pine in the black. However, it required long-term planning to buy the lumber, make the furniture, and get the new products to the stores by February of the following year. The window of sales for outdoor furniture was March to the end of May. Then the stores began slapping sale prices on their remaining outdoor furniture.

As to the 1980s' economy downturn, he reflects: "Twenty-five percent of the furniture companies across the United States had folded. With the Northwest spotted owl crisis, and the general public leaning toward fiberglass patio furniture, it was a terrible time."

From Washington, D.C., the HPM executives called Watson with the message "Our board chair says to move out. Order the redwood."

With strong misgivings, Ray ordered 20 million board feet of redwood lumber. Tension mounted in the head office when no more communication from the General Conference was forthcoming. Yet, in harmony with Seventh-day Adventist Church leadership, the Harris Pine board reluctantly voted to add the two strapped Michigan branch plants.

Meantime, Adventist Health West was approached for the needed loan. But Adventist hospitals were also in a wild state of change, and "white water" was quickly becoming the norm for that industry as well. The health-care executives turned down the request.

In November 1986 the HPM board chair took a sharp turn from his earlier meeting with HPM executives. It was his responsibility to oversee and protect the worldwide work of the Adventist Church. With Harris Pine Mills in peril, the General Conference president selected some successful businessmen "to analyze and to make sound decisions . . . a survival team."[3]

The president chose these men to advise him on the Harris Pine situation. Considered the brightest and best by church leadership, some were denominationally employed while others were wealthy executives. Not all of them understood the underlying Seventh-day Adventist mission-driven objectives.

"Selected as advisers were a publishing executive, a cement truck manufacturer, a developer and financier, the owner of a boat company, the financial officer of a hospital, the president of a health system, and the treasurer and associate treasurer of the General Conference."[4]

One morning two survival team members appeared at the mill saying they had orders from the General Conference to get the hard facts. But the questions asked by these hand-picked team members during the debriefing conferences held with HPM executives showed their lack of knowledge and indicated little understanding of this vast empire from timber to furniture. Ray Watson, who attended one meeting, recalled, "Five minutes into the conversation one survival team member declared, 'This company is broke. There is no need for further discussion.' "

Sandra Wachter-Van of the *Pacific Union Recorder* wrote:

"Unfortunately, none of the individuals selected [for this team] had experience in the lumber and furniture industries, and they apparently did not understand most aspects of lumber and furniture marketing and assessments. Perhaps that is why they believed that HPM was actually losing $50,000 per day. In reality, some of these losses were part of an annual cycle in the indoor and outdoor furniture industry.

"According to the plan approved by the board, production on indoor furniture parts was increased and stockpiled for the winter rush. Historically, in the redwood division, parts are stockpiled during the off season, from July to January. When they are purchased, they are listed as a liability, but when they are sold, they become an asset as an accounts

---

[3] *Ibid.*

[4] *Ibid.*

receivable. At the time of the survival team's study, those parts were still listed as liabilities. They had not taken into account the fact that those stockpiled parts would become assets within a few months.

"Furthermore, there were additional assets that never appeared on the team's ledgers at their accurate market value. For example, a tract of timber had been donated to the church along with the Harris Pine Mills back in 1951. At that time the tract contained an estimated 30 million board feet of lumber and was listed on the books for about $500,000. That timber has never been cut, and estimates as to the amount of lumber there today range from 80 million to 100 million board feet."[5]

Church officials were in a panic. The survival team bolstered the rumors of runaway debt, and the brethren feared the company's situation would fall upon the church. The HPM board chair called a joint meeting of the Harris Pine Mills board and the General Conference Corporation for December 3, 1986, in Washington, D.C. My father was asked to join the HPM executive team at the church's headquarters. It was clear the stakes were high.

Tuesday evening, December 2, the board chair met with four of the survival team members. "The decision to file Chapter 7 was decided that night before the large committee met. It was already a fact," Kenneth Emmerson, Harris Pine's former board chair, would reflect.

The tide had turned. No longer was Harris of Pendleton viewed as the denomination's solid financial bulwark to support and uphold, to say nothing of the $50 million poured into the student work program. Harris Pine Mills was now being viewed as a millstone. The employees of Harris Pine Mills would soon face the hurricane furiously racing toward them. December 3, 1986, proved a tragic day for the company and the church.

The moment Emmerson stepped into that meeting and saw the General Conference treasurer, who supposedly was in Africa, sitting there, "I knew something dreadful was up."

The meeting was attended by quite a few curious onlookers as well as departmental leaders and the survival team. Though sharp-edged, it had the respectability of a Senate debate, each speaker giving deference to his colleagues. But as the hours ticked by, it became apparent that the opposing forces—though uneducated in the wood products industry—were immovable. As the senior member of the Harris Pine Mills contingent, my

---

[5] *Ibid.*

father spoke for the Harris Pine group. At this point his was the only voice from the company that the General Conference president and his advisors deemed credible. His many years leading HPM through a myriad of trials, adding branch plants all along the way, and seeing the company enjoy unparalleled success, lent him credibility. His advice: keep the company. Tighten it up, and it will come through the troubled times.

In a letter to Elder Jere Patzer, president of the Upper Columbia Conference, Nagele recorded some of the ensuing events:

"The . . . two men, . . . who visited Pendleton, did not get into figures and formed worst-scenario judgments of their own with little consultation with anyone. . . . A little more in-depth study surely was in order. What about company statements and figures that stand to this day? . . . Company accountants were cast aside as amateurs, and the CPA firm [Niemi, Holland and Scott] . . . that has represented the company so well for so long as boys who don't know what they are doing.

"So the . . . estimates and guesses became the basis for the December 3 bankruptcy meeting. They were scribbled out . . . on large sheets and taped up on the wall for all to see. It was an eight-hour meeting, and no other figures were ever referred to or introduced. The General Conference president as chairman held everyone to them and cast aside the [HPM] secretary's offer to pass out the latest HPM statement. (The board secretary was also an HPM financial officer and brought the latest statements with him.) They were on the table before the chairman all day and never introduced. . . .

"The chairman declared himself neutral but soon drifted into a position clearly for the bankruptcy. So, given guesstimates . . . and the chairman buying the package, you had a formula for the split vote, and the bankruptcy was voted. Certainly not one of the church's finest hours.

"There were few who had any stomach for it, and very near half of all present never voted for it. The rest felt compelled to go along for reasons we all know too well. The HPM board resigned en masse, and those who were not GC officers were invited to leave. . . .

"In the *Adventist Review* of January 1, [1987] all the laymen were named. It is unfair to them to infer that they had any part of the consequences or the events leading up to and the decision itself. Some never attended a meeting and knew nothing about it. Only some of them were in attendance at the December bankruptcy meeting and were strangely quiet. Two of them proposed alternatives to bankruptcy and had merit. They were never picked up for discussion. By that time there were no alternatives, one would conclude by the permitted discussion."

Nagele argued to the assembly that the mill was able to withstand the downturn in wood products, could absorb the costs of redesigning furniture to better meet the modern lifestyle, and would emerge intact. There was no need to plead bankruptcy, he insisted. The company was strong enough to withstand the pressures of the current economy. "Bankruptcy is against every principle my old papa taught me," Charlie said, reflecting on how he had watched his father struggle in the timber business when he was a boy.

"The grim picture was presented to a special meeting of the board of directors called December 3 in Washington, D.C. . . . The meeting lasted an entire day," stated Sandra Wachter-Van.[6]

"Perhaps," she suggested, "if more than an 'entire day' had been devoted to thoroughly researching the problems and the options, perhaps if an adequate audit had been conducted in HPM's true assets and liabilities by someone who understood the timber and furniture industry, the church could have been spared the needless embarrassment that has followed the decision to file bankruptcy." She added, "More important, the large team of full-time and several hundred student employees of HPM could have been spared the hardship of finding themselves out of work less than three weeks before Christmas."[7]

For eight hours Nagele badgered, pleaded, cajoled, and argued. He pressed the chair of the General Conference committee and the board chair of HPM to listen to the voices of reason, to back off from the survival team's unstudied stand—bankruptcy. Finally my father said, "I simply cannot be part of a vote to take down this General Conference institution. But I'll make it easy for you. I hereby resign my position on this committee."

The resignation was accepted, along with the proffered resignations of the Harris Pine Mills executives. As the team of 11 men were exiting, replacements were being voted in, among them two survival team members. Then the board chair submitted his own resignation.

The vote to declare Chapter 7 total dissolution bankruptcy passed.

My phone rang that same evening as I stood in the kitchen stirring potato soup. It was Mother. She didn't greet me with her usual cheer. Instead,

---

[6] *Ibid.*

[7] *Ibid.*

she simply said, "It looks bad for Harris Pine Mills." Her quiet stoicism rarely betrayed hard times and bad news, and whatever trepidations she may have felt when my father left Pendleton to join the HPM team in Washington, D.C., had not caught my notice.

"What's wrong?" I asked.

In my mind I could see her at the kitchen table surrounded by the canary-yellow walls, the phone's 25-foot spiral cord allowing her to move from table to stove to sink. But she sounded drained. I knew she was sitting down.

After a 25-year absence from our Northwest roots my husband and I had just moved to Walla Walla. The move would prove providential for our parents. It was something about the catch in my mother's voice that began unraveling the incomprehensible tale for me.

"The meeting in Washington, D.C.," she said slowly, "regarding the mill's financial situation. It didn't go well."

"What happened?" Alarm bells sounded in my ears as I watched the pot bubble. What had been so urgent that my father had been summoned to accompany the HPM team? Mother hesitated. I dropped the soup ladle and slid to the floor, still cradling the phone under my chin. "Tell me," I begged.

"I can't say," she replied. "But Dad just called from a phone booth at the Dulles International Airport. He told me about the meeting and outcome and then . . ." Her voice trailed off.

My mother was a terrific listener, but that night she strained to hear him above the flight announcements and background commotion as he choked out half sentences.

I held my breath as she continued. "When he didn't say any more, I said, 'Charlie? Charlie, are you there?' He had put the phone down, and I could hear him weeping."

Very late Wednesday night, after flight delays, fog, and icy rain, a fatigued, defeated contingent of HPM executives stumbled wearily off the United Airlines flight. In a tragic coincidence of similar weather just the night before, the Pendleton Life Flight helicopter hovering seven miles from Pendleton, waiting for the fog to lift, had crashed, killing two medical attendants and the pilot. The town was already reeling.

THE MEAT-AX METHOD

CHAPTER 39

Early the next morning, December 4, Charlie embarked on a mission he dreaded—to tell Mary, the widow of Clyde Harris. He drove the few blocks to her home, pushed open her kitchen door, and greeted her nurse Alice at the sink washing up after breakfast. Mary, aged 90, had been widowed for 18 years. But unlike all his previous visits, as the affectionate son bringing news to brighten her world, he feared this news would shatter her. Strolling into the living room, he saw her wrapped in a blanket, curled up on her French settee. Her fragile osteoporotic spine telescoped her once-sturdy frame to childlike dimensions. He pulled the ottoman directly in front of her, sat down, and leaned in, searching for words. Mary trusted Charlie.

Vainly he tried to soften the blow. But shock registered in her brown eyes. Their gift to "the work," to the church, was to be *closed*? shuttered? Her small, lined face crumpled. She buried her head in the folds of the blanket and softly cried. In her letter to the *North Pacific Union Gleaner,* January 1987, Mary wrote, "We wept together."

Mercifully the added details were still unknown. The vote to close Harris Pine Mills had passed. But not in Charlie's wildest imagination could he have conjured up the scene that unfolded the next day, Friday, at the mill.

Thursday afternoon Sylvia Wells looked up to see Mr. Nagele, her former boss, walking through the office. She had been his secretary for 17 years, and she knew him well, even though he rarely came around after he retired. She was startled. "Mr. Nagele looked so changed that day," she remembered. "His face was ashen gray and haggard. He paused at my desk and said, 'Sylvia, I love my church with all my heart, but today, not the corporate church.' "

Those words would resound in her mind the next day.

Friday morning Lilly Hathaway gazed out her office window. She couldn't decide if it was raining or beginning to snow. The roads would be icy by the time she made the drive out Mission Road to her home.

Harris Pine's head office was hard at work. Even the top executives at the mill did not envision what would happen that day. "I didn't know anything until Friday morning," HPM vice president of the furniture division, Clyde Iverson, told a reporter for the *East Oregonian.*[1] As for the other employees, they knew cutbacks were in the works. Iverson had a new business plan on the table to reduce inventory and the fixed cost structure. Their order file was still strong, but waiting for the promises from General Conference headquarters had wasted months as the indecision from the top tied their hands.

Lilly looked at her watch. It was 11:00 a.m. Because it was Friday, the mill closed at 2:30 p.m. Recently furrowed brows had replaced the congenial office atmosphere, and she was silently thankful that her years of service to the company—30, if anyone was counting—were coming to an end. Her mind preoccupied with Don's decline, she had ignored the rumors floating around.

But it wasn't idle talk Lilly heard late Friday morning. One of her bosses, Hollon Fry, stuck his head in her office and said, "Lilly, you ought to park your car outside the gate right now!"

"From the look on his face, I didn't ask him any questions," Lilly says. She grabbed her purse, rushed to her private parking space behind the office, drove out the gate, left her purse in the car, and locked the doors. Then she put the keys in her pocket, went back into the building, and sat down in her office.

Moments later a loud, unfamiliar voice boomed down the hallway. "Everyone out of the building. Don't take anything with you; just file outside."

Lilly joined the group, grateful for Hollon's warning. As the snow fell on her head, she watched the evacuators nail the office doors shut.

Sylvia Wells, 34 years in the office, secretary to three successive company presidents, glanced up as four men stepped into the outer office. It was almost noon, an unusual time for unannounced dignitaries, given the mill's early closure that day. Few knew more about the mill than Sylvia. In addition to her many years at the home plant, she and her husband, Jim, had built the Shenandoah Valley Academy branch plant and spent five

---

[1] Alice Linker, "Otis Says Organization No Solution for Harris," *East Oregonian*, Dec. 7, 1986, p. 3A.

years laboring at the Virginia site before returning to Pendleton. She was planning to leave a bit early this Friday, for she needed to run some errands and to vacuum the house before sundown.

"Everyone line up against the wall. Leave your desks and belongings. Just do as we say."

"This is a new way the Adventist Church does business," said a man with a diamond ring glittering on his pinky finger.

The workers were directed to the box factory for a meeting. A few were crying. One mill vice president stood with a cluster of workers, his arm around several. A stranger to the HPM employees noticed the compassionate act, stuck his face close, and said, "You're the cause of this mess. I'd like to—"

No one was allowed back into their offices to clear out their desks. At the request of the president of the Upper Columbia Conference, Nagele enumerated some of the details of the close-down in a letter:

"Without warning on Friday . . . closure of the Pendleton plant and all branches—employees told not to come back.

"The president and vice president fired summarily. (What this accomplished is unclear.) They were also told not to return even to their offices.

"All employees' cars (bosses) were searched as they left for the last time.

"The president and vice president were given special attention—a task the watchmen almost quit before performing.

"Employees were called together in two places and were told by [the chair of the survival committee] in a cold declaration what was happening and that there were 'crooks in high places' and they would be brought down, and 'the chips will fall where they will.'

"The treatment afforded the U.S. National Bank is inexcusable and not becoming the church, culminating in the bank's request not to be asked to deal with [the survival team chair] anymore. The chairman spoke for the GC president and GC itself, the GC corporation and the Harris Foundation."

Sylvia Wells remembers the office staff filing out of the building as one of the survival team shouted from the bed of a pickup truck that the plant was shutting down: Chapter 7 bankruptcy—total and immediate dissolution.

The plunging temperature this cold, gray December day complied with the mood inside the company. Trees were dormant, the sap gone into the ground; the rain settled thick and soggy. Ice tinged against windows and bit into faces. For Harris Pine Mills and its employees, unlike nature's punctual spring to come, the winter would never be over.

Lilly recalled, "There was a big meeting out in the box factory. I wasn't there. But I heard things got pretty hairy for a while. It's probably a good thing that I can't tell you all that was said by the team that had been sent to close down the plant."

While my father kept vigil with Mary, my mother decided to find out what was taking place at Harris Pine Yards, the thriving retail store that stood at the leading corner of HPM property. The parking lot was full of customers. As she pushed open the front door, it was obvious the shutdown had not yet come to the brightly lit, cheerful atmosphere. Clerk Reggie Quinn was hustling from the lumber yard to the adding machine, totaling customers' purchases. He chatted with Mother as he worked. Earlier that morning, he told her, he'd noticed four men in business suits enter the store. They were expressionless and grim. Surely didn't act like shoppers. He shrugged and kept working. His pleasant demeanor, his laugh, but most of all his knowledge and helpfulness, drew customers like a magnet. Mother liked Reggie. She was always plotting a romance for him and frustrated with his enigmatic smile as he ignored her suggestions.

As she slipped uneasily through the aisles, glancing around at the business as usual, a purple pheasant caught her eye. Even if it was just a gaudy stencil on a salmon-pink ginger jar lamp, as an avid bird fan she paused to admire it. Then on impulse she seized it. She had a credit balance at the store and she would rescue the lamp from the netherworld of bankruptcy court. This one item would not be swallowed up by litigation and hired trustees! Forty-nine dollars and ninety-nine cents worth of fair against unfair. Reggie deducted the total from her account. Chatting lightly about the weather, he warned her to be careful up the slippery South Hill road.

For 35 years the lives of my parents had been intricately entwined with Harris Pine and Pendleton. They had accepted denominational scale pay as their part of the great Adventist movement, grateful in the knowledge that the General Conference treasury held a special "community rate" fund for Charlie's position (at his request) earmarked for missions. It was not named, but noted from the beginning, to reflect closely the rate of wood industries' executive pay.

The pheasant lamp wasn't one of Mother's best decorating choices, and she was unaware of the bankruptcy rule regarding spending with prior knowledge of what is to come. But the fluorescent bird resided thereafter beside my parents' sturdy brown HPM couches as a testament to my mother's one small protest.

An hour later when noon customers dashed toward the entrance of Harris Retail and Lumber, they bumped against a bolted door. Reggie, on the other side of the glass door, waved helplessly, mouthing the words "They took away my key! I'm sorry."

By afternoon the precipitation silently covered the unbecoming details of the day. The survival team had no denominational rulebook to consult regarding how to handle this kind of situation. In retrospect, it seems they were fearful of what to expect from the workers at Harris Pine. With their gumption roused, their imagination in full gear, the team decided force was the operative word for this company out in the Wild West.

My father would refer to the survival team member sporting a diamond ring as "the guy with a meat-ax in his back pocket."

"This is a new way the Adventist Church does business."

People in work clothes, their breath filling the factory with cold puffs of air, stood stunned, almost as if they were watching the TV report of a horrible tragedy. Banks cashed checks, even though the survival team used a wrong HPM bank account to hand out paychecks. As the first few workers stumbled in to receive their pay, banks graciously cashed checks on an increasingly overdrawn account. As the bank lines grew longer, it was clear that something terrible was in progress down at Harris Pine. The announcer at radio station KUMA said, "We should all pray."

One of the survival team ordered a secretary to call a locksmith. The man came and busied himself with the tedious job of changing every lock "to insure the contents were left on the property," the shutdown team explained.

That afternoon in Portland, the final dissolution, bankruptcy, Chapter 7, had to be affirmed and signed.

Sylvia Wells found herself mindlessly vacuuming her house in an endless loop of the connected rooms, trying to grasp what had just happened. The cozy bungalow on South Hill, built by Clyde and Mary Harris in the 1940s, was a perfect fit for the couple's Pendleton years. The close proximity to school and church was just as Clyde had envisioned.

My father received a phone call from a survival team member. Could they talk? "I am against bankruptcy," the shaken man said. "I can't do this to these people."

When Dad reminded him that he had voted in favor of bankruptcy at the December 3 meeting, the man replied, his voice breaking with emotion, "What else could I do? There was no other way."

Several months later, at the church's Spring Council of 1987, my father explained his involvement with the bankruptcy to church leadership.

"What is Charlie Nagele doing in all this?" he said. "I've asked myself the same question. I was talked into this by U.S. National Bank officials who had $24 million coming, a dozen or so of the largest creditors, the trustee, and Garwin McNeilus of cement truck manufacturing fame. Garwin's pitch did it. We had an emotional session over what had happened and how we could best salvage as much as we could from the wreckage. According to my brethren [I had to join] the Office of the Trustee 'for the good of the work.' I have no axe to grind."

"What else could I do? There was no other way" was now a hollow echo in the silent plant.

Night watchmen, chosen from employees who had been laid off, were assigned to patrol the dark, cavernous buildings and property. One bright young salesman who walked the middle of the night beat couldn't bear the silent suspension that hung everywhere. In the office, typewriters sat uncovered with half-typed letters that would never be finished.

In the finish room he saw newly designed furniture stalled halfway down the assembly line—beautiful new lines of furniture that he had helped design and for which he had been taking orders from large furniture companies. The new designs had promised to pump life into the sales of Northwest wood products despite the sagging economy.

Small piles of sawdust lay heaped underneath hastily turned-off saws. The factory was a giant silent morgue of wood and saws, paint and parts, that just hours before had blended noisily together to make up the largest wood products company in America.

The young man needed the money, but he couldn't stand being around the death of a living, vibrant plant. He quit.

NOT SO FAST!

CHAPTER 40

The fateful meeting of December 3 is history, and you can't unscramble an egg. . . . There were 40 to 50 of the General Conference's best present, plus laymen, so what more is to be said, other than it perhaps fell short of being the church's finest hour."

—Charles J. Nagele, Spring Council remarks referring to the General Conference vote for bankruptcy of Harris Pine Mills

John Mitchell was sipping his breakfast coffee Saturday morning, December 6, reading the *Oregonian* at his home in Portland, Oregon. He operated a thriving business as a CPA, Mitchell, Inc., acting as court-appointed trustee for U.S. National Bank in bankruptcy cases. Mitchell pointed out the headline "Harris Pine Declares Chapter 7 Bankruptcy" to his wife, saying with a rueful chuckle, "There's a whole lot of parts and pieces of furniture at the mill in Pendleton that are going to be sold as kindling in total dissolution."

Mitchell's interest in this church-owned business went back to a Tri-Cities Furniture auction he attended a few years before, when he bought a few pieces of Harris of Pendleton furniture. Curious about the company that had produced his newly acquired furniture, he decided to detour through Pendleton and see the plant for himself. He went into the office and asked for a tour. The friendly atmosphere impressed him, as did the whole operation. He tucked the experience into his head.

Mitchell's name wasn't next on the list as the bank began the orderly process of assigning a trustee to the HPM bankruptcy job. But on Monday morning, December 8, the assignee recused himself. The bank then called John Mitchell.

By Wednesday, December 10, John Mitchell was at the head office of Harris Pine Mills.

Nagele would tell the brethren at Spring Council 1987:

"The U.S. district court appointed a trustee and associate. In my opinion, they could not have found more knowledgeable or capable men [Mitchell and associate]."

The office staff had been told to come back on Wednesday to pick up their personal effects. As HPM vice president Ray Watson pushed through the front door of the Harris Pine office, he ran into the U.S. National Bank trustee John Mitchell and his associate leaving the building. The two men had just recently completed the Rajneesh cult bankrupty near Antelope, Oregon. Greed, malice, and murder were uncovered, scandalous details that had left both bank-assigned trustee and his assoicate with a sour attitude toward church-held properties to be litigated. "Who are you?" they asked.

"I'm Ray Watson."

"We've been looking for you. We hear you don't think the company is broke."

"The company isn't broke."

"Can you prove it?"

"Yes."

"We're inviting you to lunch," Mitchell said.

Ray's food grew cold as he talked. "We had $23 million in already-sold orders—65 percent already in the carton. Timber on the books was purposely undervalued on HPM books. It was worth many more times the $400,000 figure on the books."

"How many orders can you keep?" John Mitchell asked.

"I don't know, but I can keep almost all of them."

Watson stated his reasons for being opposed to the bankruptcy. "After an hour and a half," Watson recalls, "Mitchell asked if there were people who could confirm what I had been telling them. We came back to John's office, and one by one I called people to come and be ready to share information about their areas of management.

"At approximately 7:00 p.m. John stood up and gave us the news we had been praying for. He said, 'I want each of you to go home and call all your salesmen TONIGHT. [Tell them] that I plan to change the filing to Chapter 11 [reorganization] in the morning. Then in the morning you can contact all the branches and major customers to share the same information.'

"That night we did our calling across the United States."

As Harris Pine thanked God for John Mitchell's wisdom and his ability to listen to Watson and other voices of reason, the church's news media was absorbing quotes from the "appointed sources" as to the state of Harris Pine Mills' losses and the reason for the vote to liquidate the company.

In the church's haste to stem what their misguided figures led them to believe was a hemorrhage, the survival team (who earned the moniker "suicide squad" at Harris Pine) led the General Conference president down a wrong path. Once the decision was made to shutter HPM, the spokesperson for the church was loath to back up on the hasty facts and figures.

In addition to Charlie Nagele, John Mitchell asked Ray Watson and Charlie Fry to assist him in making sense out of what was now chaos. My father's role would continue to be a conduit between Harris Pine and the General Conference. But as he read the new items proliferating from the GC—wrong facts and figures, guesses, and even lies—he was horrified.

Ray Watson carried out his orders to call every branch plant on the evening of December 10. Mitchell's priority was to convince the church to reconsider such a harsh sentence on HPM. But the chair of the survival team railed at the bank's plan to reorganize and put the company back in business. Ray remembers that the man stormed into his office the day the court changed the filing from 7 (total dissolution) to Chapter 11 (reorganization), "accusing me of saying and doing things" that changed the court's opinion.

Ray told him, "I don't know what you are talking about. I have an assignment given to me by my boss [the bankruptcy trustee], and I want you to leave my office so I can get some work done."

MAKING MISCHIEF

CHAPTER 41

The office of the trustee developed amid our disarray with the attitude that there was a real warm place everyone could go if they [the church] didn't like what was happening. In the face of the church's own attitude and remarks, the trustee couldn't be blamed too much. However, that has all changed. Now that office has an eagerness to please the church and pay everything they have coming" (*Charles J. Nagele, 1987 Fall Council remarks*). John Mitchell's disgust at the misinformation being disseminated by the church had blocked any cordial dialogue with church leaders involved in the situation. The survival team chair claimed authority, naming himself president and chair of Harris Pine, ignoring the court's involvement and rightful takeover. He was quoted in the *East Oregonian*: "[The] chairman of the board of directors of Harris Pine Mills, Inc., said Monday that reorganization of the company 'would not have solved our problem. You have to have a reason for reorganizing, and we have no hope that reorganizing would solve anything,' he said about the company's decision to declare Chapter 7 bankruptcy."[1]

March 20, 1987, my father's letter to Jere Patzer would observe:

"The trustee proposed to change the Chapter 7 filing to Chapter 11 for reasons that are obvious. [The GC survival team chair] did not agree with this and did not join in the request. This was translated by the court and trustee as being against it. That explains why the board or . . . [the survival team chair] has no place or function. He will not accept this . . . [and still has] meetings and perpetuation of needless adversarial positions. . . . So, his mischief continues."[2]

John Mitchell told the *East Oregonian,* February 6, 1987: "'It's the best situation I've ever seen for restructuring,' Mitchell said, 'even though

[1] A. Linker, "[Survival team chair] Says Organization No Solution for Harris."

[2] Charles J. Nagele to Jere Patzer (former president of the Upper Columbia Conference of Seventh-day Adventists), Mar. 20, 1987, in author's possession.

the Seventh-day Adventist Church had become convinced otherwise.' . . . Mitchell said it appears the church's decision was based on information 'that probably wasn't entirely correct.' "[3]

In an interview with John Mitchell, the *North Pacific Gleaner* recorded him saying, "The information [church officials] relied on was not wholly accurate, and now we've found there were significant assets they didn't consider. It's very unusual to get a company in bankruptcy that has more assets than liabilities." The article continued: "Mitchell noted that many major customers have rallied and continue to buy Harris Pine Mills furniture rather than abandoning the furniture line. 'We are selling now at the same level of last year,' he said. 'We're definitely in business. This isn't somebody kidding somebody about it.' "[4]

But for a time Mitchell and the survival team chair, who supposedly represented the church's interest, remained at odds. In January 1987 my father was summoned to Portland, Oregon, for a meeting. He was instructed to meet the survival team leader in the lobby of the law office of Bullivant, Houser, and Bailey. A team of attorneys were present, lending their counsel to protect the losses of the church. By 10:00 a.m., after a two-and-a-half-hour wait, it was clear to Charlie that he had been stood up. He made his way up to the office, inquired at the secretary's desk as to why the meeting was in progress without him, and was immediately ushered in.

Charlie threw his hat on the table as he walked in, and the attorneys invited him to sit down.

"I'm not going to stay long."

"No, no, please sit down. We need your input."

"I've been waiting since 7:30 a.m. I don't intend to stay. What I have to say won't take long. I just want you to get it straight that I'm cutting off all communication with [the survival team chair]. You've lied and garbled the truth in every newspaper and every Adventist publication across the nation. I have never met a man as crass and cold-blooded as you. . . . I came down to Portland to meet with you only because I was afraid that if I let you come to Pendleton somebody would assassinate you."

The survival team chair protested, "Nagele, I was appointed to protect the office of the president . . ."

---

[3] Hal McCune, "Trustee Believes Harris Pine Will Survive," *East Oregonian*, Feb. 6, 1987, p. 1A.

[4] "Future Is Hopeful for Harris Pine," *North Pacific Union Gleaner*, Mar. 2, 1987, p. 9.

Charlie responded hotly, "I've known [the president] since before you were born, and I know that he doesn't expect you to lie for him."

The meeting was in disarray; the speechless attorneys adjourned the meeting. Later it was revealed that the legal team was planning a strategy to put the General Conference, plus costs of hiring consultants and other expenditures, before all other creditors. The plan with the request was to go before the judge that afternoon.

For the church to attempt a legal maneuver to put themselves first in the list of Harris Pine creditors was not a charitable act in my father's eyes. Charlie could not fathom the church smugly walking off with their part of indebtedness and letting the unchurched creditors absorb debts owed to them.

Bursting with indignation, Charlie reported his outrage to the General Conference president, who responded, "We will not! We will be paid last!"

But the trustee's efforts to bring rational thinking were being opposed at every turn. The mischief maker was giving the church a bad name. Trustee Mitchell caught wind of the meeting and Nagele's rant. He responded wryly, "I heard about that meeting from one of the attorneys present."

During the 1987 Spring Council, Nagele was asked to give a Harris Pine report to the brethren. In a later letter to George Thompson, vice president of the North American Division, he confessed, "It was the hardest thing I have ever done."

"I want to make it clear that my remarks today are purely my own," he began. "[The president] wants a picture of how the bankruptcy liquidation is progressing from the vantage point of the rock I stand on. I have permission to be pure, Charlie, and I told him he didn't know what he was getting into. His answer: I know exactly—and I'll risk it! It may look different to you from your rock, but that's OK. The figures are not mine but from the trustee himself. . . .

"The adversarial position we have unwittingly drifted into is, in my opinion, not the way to go. The advice 'to protect the interests of the church'—given the fact that the court and trustee's existence is to protect the creditors—it escapes me as to what that means. It gains nothing and hurts badly. We all realize that now in hindsight and are not sure how it happened.

"It seems none of the parties like it much and agree to restudy their method of communicating with one another.

"The eventual outcome of the whole liquidation procedure hinges almost in its entirety on this one man [Mitchell]. His goodwill can produce the ultimate by much effort and work, or the easy and short way of quickly taking the path of least resistance, and much less.

"The difference—and it's largely in his hands—between his dedicated efforts he now is committed to or a lesser effort could be getting the General Conference most or all of its due *or* little or nothing. It's that important!

"It is unusual to have a Chapter 11 without the stockholder or former management involved. We were set aside, and with benefit of hindsight, it may be a blessing. The board thus was set aside, and the trustee and the court replaces it.

"The stockholders, the General Conference Corporation, are for all practical purposes in the same position, except as the second-largest creditor. They can challenge actions of the trustee and court, which will trigger an automatic hearing by the court to judge the merits of the objection and rule accordingly. Further appeals would mean direct court challenges.

"Why all this as former owners? It all came with the package when Chapter 7 was filed. The trustee and the court are all-powerful by law, covering bankruptcies. The trustee makes the decision, but nothing major is done without court approval of the operating plans, sale of assets, payment to creditors, the use of liquidation money, etc. However, when it is approved, it comes down with the force of law. The only alternative by anyone outside of the trustee's office is a legal challenge or objection by a creditor, which is usually followed by hearings by the bankruptcy court.

"We don't like to hear all this, but in my judgment, and given what little I've learned, I fear we had just better get used to it and guide ourselves accordingly. We no longer own Harris Pine Mills.

"It may come as a shock, but all the figures, statements, estimates, etc., are now of little use and not a part of the figures used significantly in the Office of the Trustee. Accounting is a liquidation breed on a cash basis. The regular statement figures must be brought along in parallel for tax purposes.

"Be advised financial statements, we hope, are headed for zero—zero figures as quickly as sale of assets continues rapidly . . ."

HOLD YOUR FIRE

CHAPTER 42

Now, I have an appeal to make. To the pessimists who insist we don't know what we are doing or what time of the day it is: hold your fire; join us in our optimism. It won't cost you a cent and will be of immeasurable help to the many folks still laboring to help bring it about.

"The trustee, his associate, the creditors, employees, and even the judges read everything we give the press, and even our own papers. It may be that to this point we should have buried it all in the *Signs of the Times*, and reading our stuff as religiously as they do, we may even have made believers out of them by this time [*sic*]. Anyway, if you have a burden to prove something, you will have plenty of time to say 'I told you so' after we fall on our faces. And we will have proved your case for you.

"However, I warn you, they [trustee and associate] don't have the slightest intention of doing that. Your optimism and helping hand will be greatly appreciated. That may not earn you a cent either, but it could help put lots of money back into the church we all love so much. . . .

"There are many more assets to be put on the block. Ample to retire church obligations, given the average breaks Mr. Mitchell has received or caused to develop so far. His track record is almost a storybook one thus far, and we hope it will continue" (*Charles J. Nagele, 1987 Fall Council remarks).*[1]

The court took the reins at Harris Pine, and the Pendleton team fell in line with the court in an effort to recover as much as possible to pay all creditors. Now that court trustee Mitchell and Higgins, his associate, were heading Harris Pine Mills, all decisions had to be run through them. The HPM team and Mitchell's team struggled with the awkward chain of command. Although Mitchell and his associate were gifted at stepping into

[1] Charles J. Nagele, personal speech notes, in author's possession.

the gap and recovering assets, they still had a limited knowledge of running an organization such as Harris Pine Mills in the interim.

When Mitchell first arrived and assessed the business, he was "dazzled" (my father's word) by the possibilities for Harris Pine. He attended the furniture market in January and garnered in bigger sales than they had had in years. He felt he could resurrect the company and put it on a sound basis. As he traveled to the various branch plants and met the managers, his confidence rose at the possibility of reconstructing the company to be fully operational.

While Mitchell and his associate were away from Pendleton for stretches, the Harris Pine team were working their way through the complications, and sometimes it resulted in tongue-in-cheek give and take. The Harris Pine team often had no one to authorize operations. Nagele came into the office to oversee and answer questions, but the court-appointed team held the authority. With $6 million worth of redwood furniture still needing to be sold, Ray Watson was frustrated. His hands were tied as he must await authorization on even small decisions.

So Watson resorted to yellow Post-it notes—plastered all over Higgins's desk. When Higgins came into the office one morning after several days' absence, his work space was ablaze with the canary-hued requests. Higgins appeared in Nagele's doorway.

"What's the score?" asked Higgins, gesturing toward his office. "What kind of guy is your man Watson?"

"Well," Charlie said, "he is a thorough, well-trained man, used to making decisions and managing a well-run division. But you haven't given him any authority. When he needs a pencil, he has to ask. When he needs an eraser, he has to come to you. A quart of oil for the hyster must be run by your desk. If the truck needs gas, he has to ask."

Higgins was getting the idea. He gathered up the sheaf of yellow notes and handed the pile to Nagele. "Here," he said, catching the drift, "you decide what to do with these."

Nagele walked down the hall and handed the yellow pile to Watson. "The job's all yours. Do it your way."

Watson smiled. "Now this is more like the good old days."

In an exit interview with the trustee before Nagele left for the 1987 Spring Council, Mitchell, still feeling a little cheeky over the acrimony between him and the church's spokesman, said, "Charlie, take our warmest and best regards to your pope. Is he fallible?"

"John, I've known him for nearly 50 years, but I'm not ready to say he is infallible!"

"Well, ask him."

Higgins piped up, "You can ask my pope, and he doesn't hesitate. He says flat-out that he is infallible."

Another voice said, "Yes, but I've known your pope for 50 years too, and I think he is pushing it a bit. And I don't think he is very smart, either, or he wouldn't have set up headquarters in Italy, where the lira is no good."

Mitchell turned toward my father standing in the doorway, saying, "By the way, Charlie, I read in the paper where the guy is having financial troubles and his outfit is losing $60 million a year! Given my experience with the Adventists, if he goes bankrupt maybe I could get the job. I know you want to get back to your retirement—but would you postpone it and go to Italy to help us if we get the job?"

"John," Charlie responded, "If you get that job—that's a show I just wouldn't want to miss. Count me in!"

In an effort to quell the lingering acrimony between Mitchell and the church, the General Conference president agreed to fly to Pendleton and meet with him. My father registered his satisfaction in a letter to the president dated April 10, 1987, writing the following:

"I am most pleased by your willingness to promote a cordial and friendly personal relationship with the trustee, John Mitchell. He is likewise very pleased and wishes you to know he will enter into this very promising relationship with the world leader of the Seventh-day Adventist Church.

"He further expressed his continued dedication to a maximum effort to do the best job humanly possible for the church. He says he wants to do it, and his reputation in the field he is in calls for it, if he is to continue to enjoy the reputation he has. He sees no possible way anything could come in to embarrass you and him, but says your attitude and any concerns you have will have a great influence on his actions.

"Your willingness to be the sole spokesman for the church in the church media pleases him, and he in turn will be the sole spokesman for the public press from the trustee's office. Hopefully, just factual (no arguments) and as upbeat and guardedly positive as possible.

"Think of the favorable, healing, and quieting effect on our people everywhere when they hear you are in personal touch with the trustee. What you say and report is directly from him, a procedure, in my view, [that] could well find us in a quiet, mended state where it no longer is an issue of consequence anywhere. Especially if John's optimism prevails or comes anywhere near. As late as this morning on the telephone he said, 'Charlie, I want to make [the General Conference president] real happy, and get him all the church's money.' "[2]

Charlie reported to the brethren,

"The adversarial relationship was abandoned in favor of a friendly one. The president of the General Conference, and the trustee, are on friendly terms. These two charming personalities get along well together."

But opposing points of view and reports from an Ernst and Whinney "audit" were presented along with Nagele's Spring Council report. As recorded by Shirley Burton in the spring 1987 General Conference news and interoffice *Communiqué*:

"Two members of the accounting firm of Ernst and Whinney detailed their findings of Harris Pine Mills' business practices and statement figures over the past several years. In general, the auditors supported the overall figures of the Chapter 7 filing for bankruptcy and the earlier findings of the lay advisors. . . . The auditors found some variance from the company's own policies but no evidence of fraud. That research has been shared with the trustee and must now decide whether to accept or reject their findings [*sic*]. Charles Nagele also shared updates on the procedures of the court's trustee. Church officials will likely make no further comments in the press until such time as the trustee completes sales of the asset."[3]

Here was another hurdle. The "audit" referred to was, it turned out, a quickly gathered array of details referred to as an "audit" in church papers but was in fact entitled "Comments and Observations," by the Ernst and Whinney accounting firm. The Ernst and Whinney company pointedly entitled it "Observations," dated November 1986.

The audit company, bowing to the insistence of the survival team chair, set forth figures in this 93-page document that were wholly inaccurate and

---

[2] Charles J. Nagele to General Conference president, in author's possession.

[3] Shirley Burton, ed., *Communiqué: A Weekly Report to the General Conference Headquarters Staff*, Apr. 3, 1987.

did not reflect the true picture of the assets and liabilities of Harris Pine.

Nagele would pointedly refer to this E&W report during his Fall Council report, October 1987:

"Six months ago I stood before many of you at Spring Council. I have reviewed most of what was said, by way of a report of the activities of the Office of the Trustee under the U.S. District Court of Oregon. Its progress in liquidation of HPM assets, the payment of creditors, and our hopes and cautious optimism that most, if not all, will be paid in full: these things we set forth as it became clear to us that church estimates and figures were in serious error. That report has stood the test of these six months, and we stand by it today.

"We say that in the face of other reports later that same day [Ernst and Whinney, "Comments and Observations"] that completely contradicted our report. We have no arguments to make; time will prove their validity. Suffice it to say that in our opinion it was 'unprofessional, unethical overkill.'"[4]

Harris Pine Mills' internal audit manager, Ernest Porter, would say that the misguided report was a "combination of hectoring, asking wrong-headed questions, and lectures on accounting practices for lack of specific knowledge of such a far-flung industry." Porter added, "A primer on what an auditor does asking amiss—unskilled at the variable industry. E&W fell back on filling its pages less with observations, although it threw in the more negative observations, but more on a lecture or workbook on accounting practices in general."

Inherent in mistakes are lessons to be learned. The Lord doesn't remove all the pain, and although it is abundantly evident God put His hand over the church to "disentangle . . . from embarrassment and difficulty," as Ellen White would have said,[5] still the sting of regret remains.

Some argue a loss in the hiring of attorneys, the misguided Ernst and Whinney report, and a loss of interest on money lent. But in a letter dated June 1, 1967, to GC undertreasurer William L. Murrill, my father clarified some of the "loose ends floating around my office. Mitchell made a call on . . . [the treasurer of the General Conference] and was given a figure by one of the treasurers. It was $248,700 representing survival team expenses. It was hoped this could be put in the bankruptcy figures as due the General Conference and/or GC Corporation.

---

[4] Charles J. Nagele, personal speech notes, in author's possession.

[5] Ellen G. White, *The Desire of Ages* (Mountain View, Calif.: Pacific Press Pub. Assn., 1898), p. 329.

This figure is undoubtedly much larger by now . . . [with] the delegation plus attorneys to the Spring Council. . . . John Mitchell and . . . [the GC president] have been in contact by phone. Their friendly relationship delights me, and the new atmosphere in this office is unbelievable. The president of the 'world church' as [the GC president is referred to around here], you would think was Mitchell's associate. In another letter to his conference president, Jere Patzer, Charlie wrote, "Hold your hat, we have a good shot at a complete payoff of everyone, including the General Conference. It will be interesting to hear what 'the worst-possible-scenario boys' have to say then! [The General Conference president] is to be commended. Since Spring Council he has been a real help on a friendly basis. He has held our critics in check well."[6]

During Mitchell's peacmaking visit to General Conference headquarters, he presented Wilson with yet another dilemma the court faced. The General Conference Corporation had the right to request one of several options regarding the reinstating of the retirement benefits to salaried employees at Harris Pine. Mitchell gave a report of the cost to the church based on what the retirees were currently being paid. An alternate interpretation of the policy, from a legal standpoint, would cost the stockholders [General Conference Corporation] approximately $1 million less. Mitchell had hardly gotten the words out of his mouth when the General Conference president said, "Mr. Mitchell, I want to make it clear: we will not walk out of this process over the backs of our faithful employees."

Mitchell toured the denominational headquarters and came away from the visit with an entirely different picture of the Seventh-day Adventist Church.

As soon as he arrived back at Pendleton, Mitchell barged into Ray Watson's office. Ray was the head of the salaried employees' retirement fund, which would need $3 million to reinstate benefits owing the employees. He had met again with the church executives and was full of enthusiasm, eager to tell what had happened. With his coat slung over his shoulder and briefcase in hand, he enthused to Ray, "I have never in my life seen a more loyal bunch of employees. This is what I would have expected from a world church leader. Before I came to Harris Pine, I had just completed the process down at Antelope, Oregon, with the Rajneesh church group. And I can tell you, there was no integrity there. Your people in this organization are just the opposite. I have never worked with a more honest, trustworthy people."

[6] Charles J. Nagele to Jere Patzer, in author's possession.

I WILL RESTORE

CHAPTER 43

While John Mitchell and his team wrestled with the bankruptcy, incoming Pendleton pastor, Duane McKey, shouldered the burden of beleaguered church members. The retiring pastor, John Davidson, had not yet cleared out his desk on Friday, December 6, when he received the urgent message—"They're shutting the mill down. Please come."

McKey and his wife, Kathy, and their two children, newly arrived back in the States after 10 years in Zimbabwe, Africa, had yet to fully acquaint themselves with the close connection between the local church and the mill. It was God's providence that the two pastors had planned to work together through December as the baton passed from Davidson to McKey, for the crisis called upon both wise and caring leaders. McKey would reflect that the Sabbath of December 6 "was like a funeral. Out of the 24 church elders, 21 of them worked at the mill."

Virtually all church family units had one or more members working at Harris Pine. Through the liberal giving by mill employees and their families, the Pendleton church and Harris Junior Academy had always prospered.

Pastor Duane, who is now in full-time evangelism as ministerial director for the Southwestern Union of Seventh-day Adventists, draws a curtain over the Friday events and the ensuing days, simply saying, "It was a horrible mess." He remembers that during "the succeeding three years, 70 families moved away. It felt like Satan's curse was on the place—many, many deaths during that time," he says. But records show that the giving and tithe records of church members reached an all-time high in 1986 and 1987. "We did lots of evangelism. Two hundred fifty to 300 people were baptized during the next few years. By 1992 and 1993 our church membership rolls showed three more than before the bankruptcy."

As the people mourned, Pastor McKey pointed his flock to Joel 2:25, which says, "I will repay you for the years the locusts have eaten" (NIV).

He encouraged them to write the text on a card and place it in their Bibles, jotting down the difficulties and trials they were enduring. Then, he suggested, they ought to make notes of the blessings received—how the Lord restored. McKey reflects that again and again when he presided at funerals in those dark times, he would ask the family for the deceased's Bible, and tucked inside would be a card with Joel 2:25—and notations of losses turned to blessings.

He reflects on the conversation he had with the General Conference president a few years before he died. "We made a mistake with Harris Pine Mills," the world church leader said.

The loss to the church was its dignity. The trustee made $2 million. "Just goes to show that the trustee is human too," my father would dryly respond to a General Conference officer's complaint about the high figure.

Actual dollars lost, when it was all said and done, added up to the church's efforts to protect themselves in lawyers' fees, the voted $200,000 expenditure for the "Ernst and Whinney Comments," and other misguided avenues. And yet "the church came out smelling like a rose," as my father would say.

At the October 1987 Fall Council Nagele reported:

"Speculation that we have fallen on our faces in this disposal process is pure humbug. The market changes almost daily, and only speculation can be made as to the ultimate outcome and within educated timber knowledge limits. Without doubt, however, it will be in the millions. . . .

"A far cry from the dismal picture of 10 months ago, when it was estimated the company had $70 million in debts, and only $40 million in assets, and that the indebtedness could climb higher than that. Well, the Lord has been good and that did not prove out to be true, and today we have to guard against overoptimism. . . .

"There are many more assets to be put on the block. Ample to retire church obligations, given the average breaks Mr. Mitchell has received or caused to develop so far. His track record is almost a storybook one thus far, and we hope it will continue. . . .

"It is my opinion—and for a certainty few knowledgeable people would gainsay—that brilliant industry known as Harris Pine Mills, given to the church by dedicated members by the name of Clyde and Mary Harris, did a tremendous work and service for the church . . . .

"(1) Around $50 million was pumped into educational and student labor-assist programs. Large sums, in addition, were given to other institutions, projects, and help here at home and overseas.

"(2) Many, many thousands of young people received an education over the years who perhaps could not have any other way.

"(3) Thousands of parents were able, with this assist, to proudly educate their young people with dignity.

"(4) Help was afforded to so many educational institutions if they rightly took advantage the industry afforded them.

"(5) Over the years, thousands of Adventist lay members were employed in the 18 states in which the company operated.

"My brethren and sisters, I fear we didn't really appreciate what we had going for us until it started to falter. When the crash or tragedy, or whatever you wish to call it, came, had we thrown in the towel, given in to despair, and recovered little or nothing, Harris Pine Mills would still have been a tremendously positive venture for the 34 years I just talked about.

"If you ignored everything else, which would be unthinkable with our mission, just in terms of dollars, if we want to get that crass, we would be far, far ahead of investment or losses. . . .

"A few years ago a religious scholar somewhere shook the world by asking the question 'Is God dead?' The inference was that He just may be.

"Preachers preached about little else for a few weeks. Happily, most all came out OK and decided He wasn't dead. On December 3, 1986, for a fleeting moment I wondered about the question again.

"Fellow workers, God is very, very, very much alive. As proven by what He has, by His blessing, let us accomplish—much of which I have reported to you.

"The industry was a part of God's church and lending its best efforts to contribute to the great mission that is ours.

"That mistakes have been made is without saying. However, we are told, 'Enfeebled and defective as it may appear, the church is one object upon which God bestows in a special sense His supreme regard.'[1]

"It was true when that was written, and it's still true today."[2]

---

[1] Ellen G. White, *The Acts of the Apostles* (Mountain View, Calif.: Pacific Press Pub. Assn., 1911), p. 12.

[2] Charles J. Nagele, personal speech notes, in author's possession.

LOVE LETTERS IN THE CEDAR CHEST

CHAPTER 44

The Bustle in a House
The Morning after Death
Is solemnest of industries
Enacted upon Earth—

The Sweeping up the Heart
And putting Love away
We shall not want to use again
Until Eternity.
—*Emily Dickinson*

My parents' house was eerily silent as I let myself in that May morning of 1990. Jeep, my irascible terrier, a replica of the black, white, and tan ball of energy sent us so long ago by George Yeoman, pressed close to my heels. She was used to coming here to stay with my parents when we traveled. She and my father had a running feud on yard security. He'd close the escape routes while she sat on her haunches watching him hammer, sweat, and wire up the gap, whereupon she'd immediately find her way to freedom.

"That dog is so smart you could teach her to play the piano," my father groused one day when she had won another round in eluding backyard fence incarceration.

Today, as we moved through the quiet rooms, she pressed close to my heels. She and I had come to start packing up my parents' belongings. The drive from Walla Walla was sunny, bright, and warm. The bright green velvet of new wheat carpeted the rolling hillsides. Trees burst with leaves and blossoms, and somehow nature's energy helped ease the dark winter memories, so recent, raw, and sudden.

It began with a phone call on January 2. "Be at St. Anthony's Hospital at 8:00 a.m. tomorrow," my father told me. "My doc wants to drain a little fluid off my lungs. Nothing to worry about, really. Just been a little short of breath lately. But my heart's fine."

The highway between Walla Walla and Pendleton was icy, and I arrived at the hospital a little late, wondering why this summons out of the blue. Mother was the one who was ill. Her interstitial fibrosis of the lungs, our major concern, kept her housebound and on oxygen. But it was treatable, the doctor assured us.

As I dashed breathlessly into St. Anthony's, the greeter at the front desk directed me upstairs. Dr. Brandt, the orthopedist, stood beside Dad's bed. Dad was flexing his left arm on the overhead bar, saying, "Arm's getting weak. Need to exercise more." I saw the doctor wince as Dad tried to do sit-ups.

The internist who had tapped his lungs appeared a moment later. Dad had been X-rayed from stem to stern. "It's not good news," the doctor said. By 4:00 p.m. we would learn that tumors had invaded his lungs and brain, and one in his upper arm had eroded the bone to a sliver. By 4:00 p.m. both the internist and orthopedist surgeon had parried Dad's question, "How long?"

Neither would commit to a time frame. But I saw Dad brace himself and bend into the ill wind he read in the serious countenances before him.

The night after we transferred him to Walla Walla General Hospital, a storm broke over eastern Oregon and Washington with a wind of biblical proportions. At 2:00 a.m. we peered out our window to see our round Harris Pine redwood table lift above the patio and turn upside down midyard, levitating like a hovercraft. We endured thumps and bumps as shingles and other detritus thwacked the house. My father lay stranded on his hospital bed, paralyzed from the stroke he'd suffered the day before, blinding his right eye. He could no longer communicate except to point up—to whom he believed.

As the noisy storm continued unabated, it seemed a talisman of the struggle he had waged for so many years at Harris Pine Mills. That night thousands of acres of timber across Oregon and Washington, many of the forests he had "cruised" with Clyde and Harris Pine foresters as they approximated board feet in timber plots, were now reduced to piles of pickup sticks. The "Tollgate Blowdown" sign went up soon after, the forest service's explanation about the January 1990 disaster so apparent to travelers through the Tollgate Blue Mountains.

Was the epic storm a last swipe of an enemy who hated even the forests that signified a company favored and blessed by God? Were the millions of board feet strewn across the forests a testament that an evil force simply could not stand anything signifying a company upon which God had bestowed His favor?

We buried my father the last of January. Mother and the rest of our family tried to endure, but Mother buckled. The disease outpaced the medicine, and her ravaged body mercifully gave out a little more than six weeks later. By mid-April we placed Mother beside my father in Pendleton's Olney Cemetery.

Now I resolved to quell my emotions and dispatch the packing-up process. I paused for a moment in their short driveway on Goodwin Place, gazing at the town below and the empty spaces around Harris Pine Mills. Trucks were still carting the place away.

It had been three decades since my parents built this house overlooking the town and the mill. Before the final touches could be put on the yard, my wedding took place at the church just a block away. After the ceremony the entire congregation trooped up the hill to my parents' house. After going through the front room receiving line, they picked up slices of wedding cake and punch on the old mahogany dining room table. Then they were ushered out the sliding-glass doors to the patio, where they stood and visited for a few moments, exiting around the side of the house. Good manners demanded that we temporarily remove the garbage cans.

The view was the reason my parents decided to move from Clyde and Mary's old house on Ninth Street and why my father chose such a tiny lot. It was the last one left on South Hill's Harris Heights with the view he wanted. Despite the tiny yard and a house so squeezed it could accommodate only a single-car garage, he never regretted his choice. He got pleasure watching the mill out the sizable living room window. But late in life he could not avoid watching the dismantling process that took place for several years.

The decisions—what to keep and what must go—weighed heavily on me. The furniture in every room was proof of my father's love for Harris Pine Mills. From the front door view of the sturdy brown sofas from HPM's Upper Columbia Academy branch plant, every room brimmed with Harris of Pendleton.

But their treasures gave me pause. There was no need to call in Sotheby's to assess the china, crystal, silver, and art, for such things were missing from my parents' thrifty lives. But mementos were everywhere, brought back from General Conference overseas travels. In Africa, Dad journeyed alone between unfriendly borders, a *Review and Herald* magazine tucked into his folded arms as a familiar identifier for his mission station transporters. There was the memorable trek through South America with Elder Robert Pierson; to Indonesia and Borneo with my mother, who momentarily panicked one day, calling out, "Charlie!" when a former headhunting tribal chieftain held fast her arm. In that moment she feared the chief might try to strike a bargain to make her one of his wives as her oblivious husband disappeared down the path to the canoe.

On every trip my father preached and taught and sat down with the locals to bring additional business sensibilities into developing-world mission stations. At a remote outpost deep in Africa he encountered a Walla Walla College trained registered nurse who worked alone. Deeply moved by her singular testimony that she would spend the rest of her life and die with "her people," he held her patient steady as she chatted and stitched up a man gored by an elephant.

There was the ebony Nefertiti bust, a chip revealing it really was just white plaster of Paris. I found small, rudely carved ivory elephants and a rough wooden stick with a message from another Borneo headhunter carved into the side, pleading for a teacher. He took the request seriously, bringing it home in three pieces in his suitcase.

In the hall closet were row upon row of cardboard containers, each holding 35-millimeter slides of his travels. Without a verbal tour of the slideshow's "where in the world," context would be missing. In the tiny den, converted after my brother moved out, a small Harris Pine maple bookcase was squeezed alongside the green tweed HPM Hide-a-Bed. It was used when the guest room was taken. (Visitors usually spent only one night on it before moving on!) The bookcase held the essential family books, a distillation culled by my father's need to always clear the extraneous. However, his travel mementos seemed happily exempt from a practical, unsentimental life. All the Spirit of Prophecy books were neatly lined on the shelves. Their rich, red leather bindings spoke of the respect my parents held for Ellen White and her writings. Both of them had known and loved Willie White and his clan.

Then, unaccountably, my old childhood favorite, *The Enchanted Pillowcase,* by Josephine Cunnington Edwards, had made the cut of books to keep. In my young imagination she was unequaled as a storyteller. Well, except for Eric B. Hare. He brought his stories with him when he visited our home in the 1950s, cheerfully making himself comfortable in the old Ninth Street house basement where Dad was building and plumbing a bathroom, yet unwalled, the facilities fully visible from the stairs. Elder Hare's unharvested eyebrows danced with amusement when he told his stories and when my parents apologized for the unfinished accommodations.

The bottom shelf held the *Bedtime Stories* that included Uncle Arthur Maxwell's tales of Farmer Mack and the family of eight children, my mother the second-oldest, next to a stack of ancient copies of *Our Little Friend,* the weekly Sabbath school paper for youngsters. Mother had bound them together with shoelace ties, reminding me of her fondness for its longtime editor Ernest Lloyd. He once was admitted as a patient to St. Helena Sanitarium with unremitting hiccups. During my mother's night duty he would gently ask for mashed potatoes, the only food that seemed to help. She would slip into the darkened kitchen and heat, mash, and butter a potato. He was everlastingly grateful.

I moved to the closets. Mother didn't get rid of outdated perfectly good clothes. She merely moved them to my old bedroom she called the guest room. The closets were packed with memories.

There was the blue knit suit she wore when posing beside my brother in his cap and gown on the lawn at Walla Walla College. So many years ago. "Cum laude," she had whispered to me with pride. For sentimental reasons it was still there.

Ah, yes. There was the recent lavender suit. I remembered it from the snapshots my father had sent us when we were living in San Diego. She stood beside my cap-and-gowned father at Andrews University. He looked splendid in his purple-and-gold sash as he received an honorary doctorate.

"Don't come," Dad had said. "We'll send you pictures."

He always carried a tiny Minolta camera in his shirt pocket, snapping away wherever he went so we could share the events yet save our money. "Too expensive," he'd said when we hinted we might come anyway. Mother was already ill and breathless, her disease yet undiagnosed. But she bucked up and proved a good sport for the celebration. I smiled, remembering how the lavender of the suit contrasted nicely with her self-applied golden apricot hair color.

I kept going until their king-size bed was piled high. All around me was the Laurelcraft solid maple bedroom set from Laurelwood Academy's branch plant—Harris Pine's initiation into the hardwood market. The bedroom set was a particular source of my father's pride.

Durable, sensible, wear like iron were all terms that I had grown up with. In Dad's view it was the litmus test whether something was worth the money. He liked to show off the sturdy maple furniture, his knuckles rapping the dresser.

A memory came back to me of a long, sad tan gabardine coat my father had presented my mother soon after we moved to Portland, Oregon. "All wool, wear like iron, last forever" were the terms he used as she unwrapped her Christmas gift, to sell her on his triumphant purchase.

She tried on the Meier and Frank special, that serious gray coat. Even though they had lived in cold climates before, for some reason Portland called forth my father's attempts to cloak my mother, who had grown up in the heat of central California's farming valley. I remembered trying to cheer Mother up as she looked at herself in the mirror, all but disappearing into the straight, ankle-length dullness of his sensible purchase. She tried to cheerfully agree with him. Yes, it would last forever, but I caught the hint in her voice that she sincerely hoped it wouldn't.

There was one piece of furniture in the room that didn't have Harris Pine stamped somewhere on it. Wherever they lived, the aromatic, glossy cedar chest resided at the foot of their bed. It was my father's labor of love to the "proby" nurse at St. Helena Sanitarium. By now the old hinges were broken, and the workmanship betrayed the fact that the artisan who made it in the 1934 woodworking class was better suited for business than carpentry. Yet it gleamed under its many coats of lacquer no matter the dings and scratches of many moves through the years. I opened the lid, and the deep scent of cedar wafted up. Sliding to the floor beside my little dog, her paws on the edge of the chest, my intentions evaporated.

I had hoped to push back the memories and make practical decisions, but I could not escape the sense that the old house was closing in like a lonely mausoleum. So, letting go of the sad job at hand, I lost myself in time, tracing the early events of two young lovers who would become my parents, and lifted out one memory at a time.

The shoebox was there. I fingered the frayed ribbon, untied it, and lifted the lid. I always knew this box held my mother's heart. Letters, gray with age, the onionskin pages crackling, held fragments of my parents' history. I cradled them in my lap.

Letters flew back and forth between them during my mother's nurses' training at St. Helena Sanitarium. My father and his best friend, Frank (Francis) White, had traveled east to Madison, Tennessee, and Broadview, Illinois, having heard that the work-study programs they offered made tuition affordable. Although he was a recent honors graduate of Marysville High School, entrance into Pacific Union College required Bible credits. No exceptions.

As I breathed in the cedar, the letters and keepsakes transported me back to when their lives, even during those Depression years, spread before them bright with the future.

My father had his spies who relayed Mother's nursing class activities. The evidence was an apologetic note to Charlie written on her night shift duty. "I'm sorry, Charlie, but I so wanted to go to the corn roast that I went with Marvin. But let's not let Marvin 'bust us up.'" The shoebox didn't contain just love letters. A few gems were written by members of my mother's family. Those were postmarked Visalia, California. Her baby sister scrawled in perfect first-grade loops and curves that brother Dale had caught "the open-the-door-and-in-flu-enza." The family collaborated on a newsy four-page *Farmersville Boulevard Gazette* reporting the escapades at the farm.

Her deepest longing was to become a nurse. When it didn't seem possible after she graduated from Fresno Academy in the midst of the Great Depression, her mother took the last $75 of her modest inheritance to pay the entrance fees into the PUC nursing program. The starched-white squares lying flat in the cedar chest reminded me of watching her deftly fold them into perfect nurses' caps, ready for the black velvet band.

She met my father while pushing her tray along the hospital cafeteria dinner line, name tag pinned to her uniform. He was working in food service, had recently been baptized, and hoped to attend PUC in the fall. While splitting wood one day on Howell Mountain he had learned her name from one of the guys he worked with—her Fresno Academy senior class boyfriend.

"She'll make someone a fine wife," the former boyfriend remarked casually, and Charlie remembered her name. And suddenly he saw *her*, pushing her tray along, name tag in place.

"Hi, what's your name?"

Can't he read? she thought.

She would soon learn about the hardworking family on the hill. For Charlie, coming to Howell Mountain wasn't as appealing as it had been for my mother. Dread, fear, and resentment underscored his family's relocation. Charlie's mother made the decision during her husband John's pneumonia delirium one muggy summer afternoon, some 90 miles to the north.

The quick-freeze business and dairy store in Marysville had been his father's livelihood since selling his partnership in a northern California logging operation where Charlie had spent his childhood tagging his father in the woods. From dashing in and out of the walk-in freezer in the heat of midsummer, John developed pleurisy, an inflammation in the lining of his lungs.

He had arrived by ambulance at St. Helena Hospital, unconscious. Marie, Charlie's mother, was determined to introduce her balky husband, even against his will, to her newfound faith, the Adventists. It was she who had befriended a frail little man who had wandered into the dairy store for a nickel glass of buttermilk to soothe his ulcer. While selling Adventist books door to door in the Pacific Northwest he'd dream of a family in northern California who needed his Bible studies. Wamsley was grateful to find the Wild Rose Dairy store and became a steady customer.

Marie's heavy Swiss accent—her journey at 18 to America through Ellis Island, leaving behind her Jesuit-trained father, her Protestant mother, and a stern lifestyle as the eldest of seven children—informed her life. As the gentleman drank his buttermilk, he began unfolding a kinder view of God to this hardworking woman. Soon she was baptized and invited a few women from church to her living room to pray for her family. Daughter Hedy was soon baptized. But father John didn't like stumbling over his wife's kneeling friends. He did a little drinking, a little gambling, and he found his wife's religion and piety distasteful. Teenage Charlie agreed with his dad. Religion had gotten to be a heavy subject in their home. The family split their loyalties and work schedules to avoid one another.

But with John unconscious, Marie saw her chance. She directed the ambulance to deliver him to the St. Helena Sanitarium. But the plan backfired. When his fever delirium abated and he discovered he was in an Adventist hospital, John demanded his pants and prepared to leave. The commotion alerted an ambulatory patient convalescing at the hospital. He, like Wamsley, was an ailing colporteur. God bless the literature evangelists! With unassuming charm he looked in upon the scene of loud disgruntlement. "I'm leaving," John rasped.

The colporteur got the picture. He sat down and provided some reasons that the patient, who was coughing blood, shouldn't leave. "John, you're too sick to let these folks upset you by their strange beliefs. Even if you're here without your consent, you need the best medical help available, and this hospital has wonderful doctors and nurses. I recommend you tell the doctors you're not leaving until they get you well."

John slumped back into bed in great pain and regarded the stranger's advice. When the doctor appeared, John told him that he wasn't leaving until the place made him well.

A friendship developed—between the ambulatory patient and the sick man who faced an uphill battle before the introduction of antibiotics. Again and again the subject of religion came up, mainly because the Marysville businessman was so exercised over the subject. But his new friend read the Bible in ways that opened a new understanding to John. He gave him a Bible that John hid under the mattress, lest Marie get the idea he might have a spark of interest. His friend pointed to texts by page and number, and truths became clear in light of the 66 books. As his new friend spelled out God's great love, the plan of salvation, and the beauty of the Sabbath, the ailing man grasped a new picture of God, so different from the arbitrary God he believed had driven a wedge into his family.

During the 13 months of the hospital stay Charlie, driving a truck route for his father's business partner, visited his dad on Friday evenings. Winding up the narrow road to the sanitarium, he'd leave the engine idling while he dashed up the back stairs to his father's room, unaware he was raising the ire of the Sabbathkeepers.

Father and son shared a strong bond. And now as John surreptitiously pulled the Bible from under the mattress, he recounted what he was learning. Theology was missing. But one thing Charlie didn't miss was the passion this new view of Mother's religion fueled in his father. "Charlie, we can't deny it. Mother is right," he would say.

Week after week, the Friday night page-and-number studies continued until Charlie's father put the question to his son: "Will you join me in keeping the Sabbath?"

Charlie said yes but didn't know the resistance his boss, John's friend and business partner, would throw up. When Charlie summoned the courage to ask, he was met with derision. "You tell John he's lost his mind. If you want Saturdays off, get another job!"

Grinding through the gears up the hill the next Friday night, Charlie battled his inner turmoil. How could he tell his ailing father that he simply couldn't keep Sabbath and also pay the hospital bills? Through those hours as he drove the curves, he wrestled. Finally he made a decision: for his dad's sake, even as the hospital bills continued to mount, he would do it.

And so it was that he traded truck driving to become assistant cook, beginning on weekends. (Sabbath workers received no pay.) But his dad wasn't through. "Charlie, I'm going to be baptized on Sabbath. I want you to join me."

But Charlie wasn't sure he wanted to go that far, even for his ailing dad.

Fun-loving, honest, hardworking Frank White noticed his friend Charlie's misgivings. The guys had planned a camping trip to Clear Lake on Charlie's weekend off. It was a perfect excuse for Charlie to miss the baptism. But Frank sensed the turmoil and quandary going on in his friend, despite the wonderful time they were having. When Frank broke camp at noon that Saturday, Charlie knew what was up. They arrived at the old St. Helena swimming pool behind the hospital in time for the 4:00 baptism.

Despite 13 months of the best doctors and treatments, the pleurisy finally won. A group of nursing students were welcoming the Sabbath by strolling through the halls singing, "Let us sing a song that will cheer us by the way, in a little while we're going home." As the sweet strains swept down the hallway, John slipped away.

"And no tears will fall in that city bright and fair; in a little while we're going home."

In the cedar chest I touched a small worn Bible. On the flyleaf was inscribed "John Nagele." I held the secreted page-and-number scriptures that had convinced my grandfather to trust God wholeheartedly.

At the back of the St. Helena graveyard is a simple plaque, J. Nagele. It's a crude stone, probably for identification purposes only. A dirt path separates it from the potter's field beyond. It would take the three bereft family members several years of hard toil to pay the hospital debt. They simply could not afford a gravestone.

My father's education about Adventism was more clearly defined and forged on the front porch of Rosehaven, Willie White's home, which was located directly behind that of Elmshaven, the home of his mother, Ellen White.

Francis, Willie White's youngest son, often brought his pal Charlie to his family home, where they'd find Willie sitting on the porch in his rocking chair. While the fellows slouched against the pillars, they listened to Willie pour out stories about Mother White and the church. So it was that Charlie's theological training was of a most unusual variety. Francis, Charlie, and Willie—and stories of Adventism's beginning from Willie White, who had been there and was a vital part of its early years. As Christ's disciples could testify what *they* had seen and heard at the very beginning of Christianity, so Willie testified of God's providences at the beginning of this church. Charlie held sacred the writings of Ellen White, her life and ministry. On that porch he developed a lifelong trust in the Adventist Church, how it came into existence, and how and why its beliefs were forged. Nothing could shake his loyalty to the church he loved so well.

After my parents were married, their first home was the third-floor apartment at Rosehaven, Willie White's home.

As Charlie was finishing his business education and looking for a job, my mother was caring for a patient in Oakland General Hospital by the name of Rudolf Spreckels. He was heir of the Spreckel sugar dynasty. She introduced him to her husband, and in this tall, dark-haired young man Mr. Spreckels perceived a keen business acumen. At the time my parents had moved to an apartment in the back of a funeral home, for Charlie had been toying with becoming an undertaker. He took night calls, and my mother assisted in funerals as a nurse. Now Mr. Spreckels offered my father a position in the sugar company at a salary that dazzled the young couple. Could they turn it down?

Perhaps it was no coincidence that just then the Northern California Conference needed a cashier-bookkeeper and offered Charlie the job. Would it be Spreckels' offer to groom him for advancement in the wider business world, or the lowly job with the Adventists? Mother pressed for the bookkeeping job, fearing the temptations that would come with the leap into high places. What to do?

The cedar chest held the answer.

I fingered two sturdy olive wool Master Comrade sashes, his and hers. Both were heavy with the felt badges signifying earned honors—he as junior camp director and she as the nurse—during the summers when Dad worked at the Northern California Conference. The honors were appliquéd with appropriate symbols, a bandage signifying first aid and so on. My parents had earned the achievement of Master Comrade. (During the cold war years that title was changed to Master Guide.)

Two varnished wood knots, safety pins glued to their backs, displayed the wood-burned carving "Tahoe Jr. Camp, 1943." And suddenly my mind filled with memories of the brisk air, giant pines, picnic tables, and breakfast outside under the trees. Again I heard the big voices of hearty staff members. There I sat, a reluctant 3-year-old, staring sleepily into an impossibly large bowl of steamy oatmeal, my stomach rebelling.

So that's where I learned that ditty, I thought. I'd even occasionally used it on my own balky little boys. The strong, brave voices urging me on, echoed again—belting out the "Are You Sleeping, Brother John?" tune to their own lyrics . . .

"Down that breakfast,
Down that breakfast,
Grumblers hush,
Grumblers hush;
We must grow up healthy,
We must grow up healthy,
Eat that mush,
Eat that mush!"

I fingered the two olive-and-tan sashes reminding me of the morning serenade. Were these sashes the source of my father's love for drab gabardine? My fingers traced the honor badges as I wondered how their journey had finally taken them to a corner of eastern Oregon and a lumber and furniture industry.

Yet it fit with my father's upbringing. Born to sturdy Swiss parents, both newly immigrated, their work ethic demanded much of each family member. His father's northern California logging venture—proof of which resides in an old family album, my father, just a kid, perched upon an enormous log in his short pants and knee socks. Years later something resonated in Charlie when Clyde Harris and church leadership tapped him for Harris Pine Mills. It wasn't such a stretch, after all.

Suddenly out of the silence the doorbell rang. Jeep barked. I ran my sleeve over my face, wiping my eyes. I'd lost track of time and place. On the front porch was a woman with a smile so huge, so welcome, that I wept all over again. It was Ruth, married to a genius of a man, Rocklyn Rouse, who had interrupted his Ph.D. with mission service in India, then come to Harris Pine for the rest of his denominational career. He'd devised ways to modernize the powerhouse and created a better heating mechanism for the mill. Ruth, a wonderful woman, brilliant in her own right, a lover of people and books, was always seeking to brighten others and lift their load. She bolstered my mother's spirits when my father was far away on extended business and mission trips. That Ruth!

"I saw your car in the driveway and thought you could use a hug," she said. Ah, the joy! I escorted her to the bedroom where the bed was piled high. "What can I do with all these things?" I asked as I held up a parkalike full-length down coat.

"Oh, I have a friend," said Ruth, "who doesn't have a winter coat. She'd treasure that." Delighted, I handed it over.

For every piece of clothing Ruth knew just the person who needed it. We loaded her dusty-tan Datsun to the ceiling. She had a way of turning sad into happy by knowing just where the needs were.

I sang as I drove back to Walla Walla. My questions, the whys, the doubts, the losses, had evaporated. A certainty replaced all the questions. I no longer wondered about this journey through life that my parents had taken, bringing me along to witness the unfolding wonder of it, getting to know and love this place and the people who surrounded the story of Harris Pine Mills. Beyond any doubt, the Lord's love and care were written all over it. The Lord loved the givers and increased what was in their hands to bless His people. A place with a friend named Ruth. Priceless!

MOVING ON

EPILOGUE

Mary Harris lived her last years with niece Margaret deRomanett and family in Loma Linda. She struggled against old age; her losses, including the mill; and how to deal with her own considerable assets when she died. Even though Clyde had arranged for all their money to go to the General Conference of Seventh-day Adventists, now she became reluctant to follow the plan he'd set down so many years before. Her grandniece, Linda, a physician, spent hours at her bedside consoling her as the tears flowed, tending to her, speaking gentle words of comfort. Aunty knew about Linda's zeal for Adventist radio ministry. Mary had spanned the generations from the admonition of her mother, Susannah, that she must do as her husband said to the age of a woman's right to declare and assert herself; Mary, now pushing 100, became a modern woman. She changed her will.

And so it was that Mary created the Mary V. Harris Foundation, all proceeds going directly into the purchase of radio stations and gospel preaching across the United States. *Family First* has become a series of stations dedicated solely to sharing God's love through the airwaves. To date, more than 500 baptisms have resulted—individuals from all walks of life who have unexpectedly entered an Adventist church and requested baptism. Mary died in 1996 at the age of 99 and was laid to rest beside her beloved Clyde in the Milton-Freewater Cemetery. Angels must hover near, anxious to bear her babies to her arms at the resurrection.

It's been more than 60 years since the night our family Packard glided past the Harrises' home and I gazed up at the grand white house on Goodwin Avenue. It was the moment for me when *Lock, Stock, and Boards* began to magically unfold. The century has come and gone, but the story lives on. The couple who lived there left an indelible mark on the Seventh-day Adventist Church. Their house still sits on the brow of South Hill, although it seems so much smaller than in my 1951 memory. The street crowds the house, and the aging rock wall teeters toward the sidewalk. Yet a few years ago I began to wonder—was there any last shred that might bind the story to our hearts?

It was novelist Thomas Wolfe who penned a work entitled *You Can't Go Home Again,* and perhaps it's true. But we can look back from whence we have come. I wondered, *Was everything gone?* Was there nothing left to hark back to the 1950s nostalgia of the gift fragrant with pine shavings, memories of clatter, and the happy commotion of the once-thriving business, yet so much more?

I had no clear idea of what I was looking for. But I knew that good, decent people made a living there and believed their toil was for a higher purpose—the gospel commission in work boots, if you will. Those moments at the General Conference sessions every few years and the annual Spring and Fall Councils, where my father would speak briefly of the Lord blessing the industry, then hand the General Conference treasurer a check for $100,000 above the continual adding of branch plants and other church contributions.

I remember particularly the 1954 General Conference session when a ripple of awe swept over the audience and the San Francisco auditorium rang with applause.

Harris Pine's steam whistle still echoes in my memory. My husband, Rodney, and I, both with roots in the mill's rich history, returned to Pendleton some 10 years after the mill's demise on a tour to remind ourselves of our heritage. We crossed the undulating hills, fields of wheat just harvested, newly cut alfalfa, endless acres of harvested sweet peas, and golden expanses spreading clear to the hazy, distant mountains.

With nagging trepidation we wondered if our journey would call up depressing memories, evidence that we're still shackled to this world, however much we dream of a better place. Changes aren't always better, just different.

My husband's parents are buried at Mount Hope Cemetery in College Place, Washington. My parents are on the hillside in Pendleton's Olney Cemetery. Their tombstone echoes the theme of my father's last sermon, taken from the hymn "The Way of the Cross Leads Home." Their tidy green house, built in 1959 on South Hill, was bought by a semiretired associate pastor of the Pendleton Adventist Church.

We rolled over the old viaduct into town and encountered the familiar east-west one-way signs for Court Street leading to the mill site and Dorian, named for that brave Native American woman of the 1800s who survived a fierce winter alone and homeless with two tiny children after her French husband's murder.

We went west on Court Street, past Hamley's, the legendary saddle maker, the sign in the window declaring, "We've been right here for 100 years." We decided to check the place out. It was like entering a time machine. They still make and sell ornate hand-tooled saddles for cowboys. They do a brisk business in blankets and tack gear, mysterious to greenhorns yet essential to modern cowpokes.

Traveling through town, we passed the Round-Up stadium's white balustrades that hug the town's main corridor so closely that I wonder what young lawyer Roy Raley, the 1920 first Round-Up association president, was thinking.

After negotiating the steep curves of the Blue Mountain's Cabbage Hill, Harris Pine logging trucks roared down this narrow boulevard, going to the log pond. The Blues, as the mountains are known locally, earned their stern reputation as far back as 1836. The first missionaries to the area, Whitman and Spalding, were forced to abandon the last remnants of Spalding's little wagon in order to negotiate the shale precipices before they finally reached the valley floor. In modern times not a few loggers, their brakes too hot, realized they were picking up speed and had to slow their rigs by sideswiping the stone embankments before the sandy off-ramps were built for runaway logging rigs.

In the summer the ubiquitous pea vine trucks added to the town's character. Peas, a staple crop in eastern Oregon, headed for the canneries. The vines, in various stages of fermentation, a cow's delight, were heaped high onto trucks, tendrils floating up and away as they roared through the center of town. The truckers seemed to take perverse glee in sullying the main boulevards with the sweet stink.

We neared the old HPM site, bumping over the railroad tracks, across Highway 84, the corridor to Portland, where so many kids after a summer at HPM left for Laurelwood Academy. But it was a toss-up whether eleventh-grade Pendleton kids would head for Laurelwood or Upper Columbia Academy near Spokane.

Commotion still reigns on the corner of Twentieth and Court streets. Once it was Harris Building Supply. Now it's a four-way modern maze of neon signs. What once was control central for an industry encompassing standing tree to furniture has been covered with asphalt. Walmart hunkers down and beckons us in. Sentimental HPM workers Jim and Sylvia Wells led a petition to Walmart executives for permission to hang a plaque and pictures on the entry wall, "In Memory of Harris of Pendleton." The mill had modernized its name, but to those of us who lived and worked there, it would always be Harris Pine. The entry wall display seemed to be all that remained. And yet . . .

The Harris site had been scrapped and redesigned into a Taco Bell and Walmart. But we passed up the burritos, dodged Walmart shoppers and carts, and drove to the lot behind. Looking back, who would have predicted that the northwest corner of Pendleton, so long the church's pride and joy, would one day differ so dramatically from what it used to be. But like the landscape, we too have changed.

It was like another country behind the biggest retailer in America. The only sign of life was a delivery truck backed up to a loading dock. Acres of tumbleweeds, strewn trash, and swirling dirt decorated the barren landscape. That is, except for hulking cement pylons jutting out of the rubble of broken slabs, dirt, and gravel. My initial reaction was that the cleanup crew grew tired of tearing down and hauling away the remains, and simply quit.

We turned the car's engine off and sat in silence, staring into the weeds and trash. Like the tangle of our lives, Walmart has a comely face; the messy, unkempt detritus is out of sight.

I pointed at the pylon. "What is that?"

My husband's memory went back further than mine and included details that had missed me completely. "I think it's a foundation for the boiler that provided steam power for the mill," he replied after some thought. The massive heating system was centered in one place on the HPM site. Fueled by sawdust, it provided steam power for the sawmill and furniture factory and propelled an electrical generator that provided electricity for HPM and also for a section of the city.

I knew that story by heart. It had been told again and again in Adventist pulpits and periodicals as an object lesson in faith.

"Interesting it's still here," he mused. Long shadows played on the jumbled, desolate terrain as the thin autumn sun dipped low. We were parked facing a curious large cement slab about 30 feet away, perfectly intact, long scored metal tracks defining parallel patterns across the surface.

An insistent nudge urged me to ask, "Is that a part of the dry kiln slab?"

"It has to be!" he said. "That's where the steel rails guided the loads into the kiln." We looked at each other, our thoughts colliding. I remembered the 1973 *Review and Herald* article written by General Conference president Robert Pierson. He wrote:

"Ever see God do the impossible? I saw Him do it this weekend here in Pendleton, Oregon. You see, in the mother plant of the Harris Pine Mills organization, huge boilers and kilns capable of drying a million feet of green lumber a week are very important installations in the milling and furniture business. . . .

"State boiler inspectors have made it clear to the Harris Pine Mills administrations that boilers just cannot be shut down every week. 'Expansion and contraction from resultant severe temperature changes will loosen firebrick mortar, damage water tubes, and seam welds,' they explained.

"Drying kilns can't be shut down while the lumber is being dried, for it must meet exacting standards. Any deviations result in downgrading of surface structure. Case hardening and warp in the resaw and other manufacturing processes will occur if the drying process is not constant. 'Never interrupt a drying cycle once started,' said the experts. 'This will prevent proper drying to exacting standards.'

"It can't be done, but at Harris Pine Mills it *is* done every week. Fires are shut off in both the boiler and kiln every Friday evening before sunset and turned on again after sundown Saturday night. Boilers and kilns alike keep the Sabbath at Harris Pine Mills."[1]

Suddenly it was clear we were staring at the modern-day equivalent of the prophet Samuel's monument. "He named it Ebenezer, saying, 'Thus far the Lord has helped us' " (1 Sam. 7:12, NIV).

As we grew up we took it for granted the dry kilns would be shut down on the Sabbath. Clyde Harris began the tradition. Sabbathkeeping was a part of who Clyde was. The boiler that powered the mill was switched off Friday evening at sundown. Mr. Gesele, an early superintendent of the power plant and dry kilns for many years, followed by others, including Rocklyn Rouse, were unusually gifted men with an innate knowledge of engineering. They simply implemented ways to carry out Clyde's original instructions to shut the power plant down on Friday evenings.

It was an interesting fact but not the big deal that scientists and engineers

---

[1] Robert H. Pierson, "God Does the Impossible," *Review and Herald*, June 14, 1973, p. 2.

made of the two weekly occurrences. The experts came and went through the years installing, observing, and sometimes fixing those two integral parts of Harris Pine Mills. When a million-dollar boiler was bought in 1961, factory installers made it clear that the manufacturer would not guarantee the new heating system if operated on an every-seventh-day-shutdown method. Harris Pine Mills took its chances—no guarantees. Furthermore, lumber experts flatly declared that the drying cycle of green lumber could not be randomly interrupted without producing "snaky" lumber. Yet no one paid much attention to the straight lumber that continued to roll out of the kiln no matter what day of the week it went in. For 24 hours each week, Harris Pine had always acted very much out of the order of wise business procedures.

My husband and I were both lost in thought as the fall afternoon sun faded. Here was evidence that God, who watches over things both big small, had not been obliterated. Mute testimony was starkly obvious in the tumbleweeds and trash that Harris Pine Mills was not just a curious footnote in Seventh-day Adventist history.

It rang clear that in the midst of our messy, tangled lives, with the world spinning out of control, overarching everything is a plan. God's wisdom and love transcends our puny works, politics, and the messes we make. He is in charge and always has been. The Ebenezer was there, trash and weeds notwithstanding. Time after time God provided a path to follow when the future looked bleak. If we look closely, we'll see Him at work in our own everyday lives.

In a copy of a two-page 1963 letter, sent simultaneously to my father and to Elders R. R. Figuhr, T. Carcich, and C. L. Torrey of the General Conference, auditor E. L. Becker, CPA, head of the General Conference Auditing Department, displays his feelings about Harris Pine Mills' real worth:

"While we all feel happy at the prosperity of the Harris Pine Mill operation and grateful for the continuing blessing of God in this operation, it has been my personal conviction that the greater value represented by this wonderful gift of Brother and Sister Harris is in the aid it is giving to our educational program in many places around the United States. There is a value that cannot be measured solely in dollars, and it appears nowhere in the financial reports of the Harris Pine Mills. In the more completely and accurately kept records of the angels, however, I am sure this imponderable value glows with the luster of heaven."[2]

Cue the "Hallelujah Chorus."

---

[2] May 15, 1963, in author's possession.